István Bíró

# Examples for practice in mechanical motion for mechanical engineers

István Bíró

# Examples for practice in mechanical motion for mechanical engineers

LAP LAMBERT Academic Publishing

**Impressum / Imprint**

Bibliografische Information der Deutschen Nationalbibliothek: Die Deutsche Nationalbibliothek verzeichnet diese Publikation in der Deutschen Nationalbibliografie; detaillierte bibliografische Daten sind im Internet über http://dnb.d-nb.de abrufbar.

Alle in diesem Buch genannten Marken und Produktnamen unterliegen warenzeichen-, marken- oder patentrechtlichem Schutz bzw. sind Warenzeichen oder eingetragene Warenzeichen der jeweiligen Inhaber. Die Wiedergabe von Marken, Produktnamen, Gebrauchsnamen, Handelsnamen, Warenbezeichnungen u.s.w. in diesem Werk berechtigt auch ohne besondere Kennzeichnung nicht zu der Annahme, dass solche Namen im Sinne der Warenzeichen- und Markenschutzgesetzgebung als frei zu betrachten wären und daher von jedermann benutzt werden dürften.

Bibliographic information published by the Deutsche Nationalbibliothek: The Deutsche Nationalbibliothek lists this publication in the Deutsche Nationalbibliografie; detailed bibliographic data are available in the Internet at http://dnb.d-nb.de.

Any brand names and product names mentioned in this book are subject to trademark, brand or patent protection and are trademarks or registered trademarks of their respective holders. The use of brand names, product names, common names, trade names, product descriptions etc. even without a particular marking in this work is in no way to be construed to mean that such names may be regarded as unrestricted in respect of trademark and brand protection legislation and could thus be used by anyone.

Coverbild / Cover image: www.ingimage.com

Verlag / Publisher:
LAP LAMBERT Academic Publishing
ist ein Imprint der / is a trademark of
OmniScriptum GmbH & Co. KG
Heinrich-Böcking-Str. 6-8, 66121 Saarbrücken, Deutschland / Germany
Email: info@lap-publishing.com

Herstellung: siehe letzte Seite /
Printed at: see last page
**ISBN: 978-3-8433-5926-9**

# Examples

# for practice in mechanical motion

# for mechanical engineers

**István Bíró**

**Lambert Academic Publishing**

**2015**

# Content

Introduction ........................................................................................................... 5

1. Kinematical synthesis of some mechanisms ........................................................ 7

   1.1. Path generation synthesis .............................................................................. 8

   1.2. Function generation synthesis ...................................................................... 21

2. Some dynamic examples in mechanical engineering ........................................ 31

   2.1. Nonlinear mechanical vibration .................................................................. 31

     2.1.1. Nonlinear mechanical free vibration ...................................................... 31

     2.1.2. Nonlinear damped free vibration ............................................................ 33

     2.1.3. Comparison of numerical and analytical solution – in linear case ........... 36

     2.1.4 Example for geometrical nonlinearity .................................................... 37

   2.2. Natural frequencies of torsional vibration .................................................. 40

     2.2.1. Single degree-of-freedom oscillation .................................................... 40

     2.2.2. Multi degree-of-freedom torsional oscillation ...................................... 42

   2.3. Critical shaft speed ..................................................................................... 49

References ........................................................................................................... 55

# Introduction

In this book some worked out kinematical and kinetic examples can be found. These examples can be useful not only for practicing constructor mechanical engineers but also for mechanical engineer students.

Applied methods are relatively easy to use. For solving equation-systems the Solver of MS Excel is suitable tool. In kinetic chapter there are a couple of differential-equations to be solved using simple numerical method which is presented in this chapter as well.

In first chapter some kinematical examples in the field of synthesis of planar mechanisms can be shown. By presentation of a few examples concerning four-bar and slider-crank mechanisms this chapter covers path generation synthesis and function generation synthesis.

In case of path generation synthesis there are a series of for example from technological point of view important points. There is a specific point on the coupler which draws a trajectory that passes through these points. The task is to determine the design parameters of the structure.

Different transmission functions are the basis of function generation synthesis. During the synthesis of planar linkages it is necessary to determine the design parameters of the structure. As a result of synthesis it can be obtained an approximation of an expected function between input and output links.

In second chapter there are a couple of application of numerical method presented by examples of single degree-of-freedom nonlinear mechanical vibration and multi degree-of-freedom linear torsional vibration of beams.

In case of construction of oscillating systems the most important parameters are their natural frequencies. Because of phenomena of resonance the avoidance of them is definitely needed.

The last subchapter is about the critical shaft speed. It is impossible perfectly to balance discs secured on elastic shafts for this reason the rotating shaft becomes bent by inertial forces. At critical shaft speed the unbalance and the deformation of the shaft could tend to infinite. It is needed for the constructor to take into consideration of this phenomenon.

# 1. Kinematical synthesis of some mechanisms

Kinematical synthesis of planar and spatial mechanisms and different optimization methods lead directly to a design that fulfills all expected requirements or at least is the optimum one considering some desired conditions. Generally the problems of kinematical synthesis are treated geometrically by the methods applied in this field.

The tasks are the optimization of mechanisms from different point of views. The expected conditions can be complied by modifying of certain geometrical parameters such as lengths of the rigid bodies, positions of no moveable kinematical joints, angles between axes of different joints, and so on.

Some program packages for optimal synthesis have been developed in recent decade. These are applied to many different types of planar and spatial mechanisms. Generally these programs are based on different types of numerical methods for optimization seeking the optimal solution with a minimum level of objective function.

In this book the mechanisms are described by natural coordinates. Using of natural coordinates leads to a simple system of constraint equations. Therefore the applied method is also simple and efficient.

The kinematical synthesis of multibody systems is basically geometrical problem. In the frame of this book presented examples about the field of kinematical synthesis can be separated in two groups such as

    ✓ path generation synthesis and

    ✓ function generation synthesis.

An industrial application of path generation synthesis can be seen in Fig 1/1.

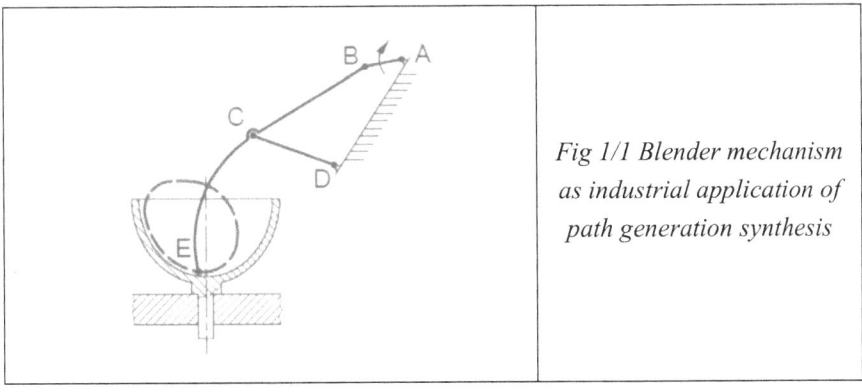

*Fig 1/1 Blender mechanism as industrial application of path generation synthesis*

## 1.1. Path generation synthesis

In this book the path generation synthesis is presented by examples of a planar four-bar and slider-crank mechanism. There is a specific point on the coupler which draws a trajectory that passes through a series of expected (predefined) points or at least goes as close to them as possible.

Applying natural coordinates leads to the simplest numerical treating of mechanical systems because the geometrical definition of the mechanisms and the interpretation of the computed results are relative simple. The attribute "natural" means: coordinates which describe the mechanical system on the basis of the characteristics of the different types of kinematical joints.

The creation of constraint equations is also simple because they are line or quadratic equations usually without trigonometrically expressions. Using natural coordinates the position of a rigid body can be determined by the position of two or more points connecting to the body and the components of unit vector connecting to two or more bodies.

Since the points and unit vectors can be placed in the kinematical joints they can be divided along the kinematical chain. In this way the number of the coordinates of the mechanisms reduces and for this reason the number of the unknown variables in the equation system reduces as well. It follows from this that in such cases less mathematical operations are necessary during the design process.

Method presented in this subchapter may be easily applicable to other planar and spatial movable structures. Generally speaking the following steps have to be taken into consideration in order to find the optimum solution for the investigated mechanism:

- ✓ Choose the topology of the structure. As a result of this step the engineer should know the number of rigid bodies with their topological description and the types of the kinematical joints.
- ✓ Determine the design parameters. They can be in general geometrical dimensions of the bodies, angles between axes of different kinematical joints, and so on.
- ✓ Define the design requirements.
- ✓ Create the geometrical and the functional constraint equations.
- ✓ In case of necessity define and minimize the objective function.

**Example 1/1** – *Kinematic synthesis of a planar four-bar mechanism*

In Fig 1/2 the sketch of a four-bar mechanism is shown. Part marked by *2* is the driving one. We have *5* predefined points on the trajectory of point *E* on the coupler. Let us call them set of design points ($P_1$, $P_2$, ...., $P_5$).

Coordinates of predefined points, *A* and *D* are:

|  | $P_1$ | $P_2$ | $P_3$ | $P_4$ | $P_5$ | $A$ | $D$ |
|---|---|---|---|---|---|---|---|
| x, m | 0,100 | 0,140 | -0,260 | -0,445 | -0,345 | 0 | 1,200 |
| y, m | 0,700 | 1,170 | 0,940 | 0,483 | 0,262 | 0 | 0 |

The task is the following: determine the dimensions (design parameters) for point E in order to pass through the predefined points. (Points *A* and *D* cannot be moved, $L = \overline{AD}$ ).

> **Important remark:** predefined points are given in coordinate system *xy* (Fig 1/2). Results depend strongly on position and orientation of applied coordinate system.

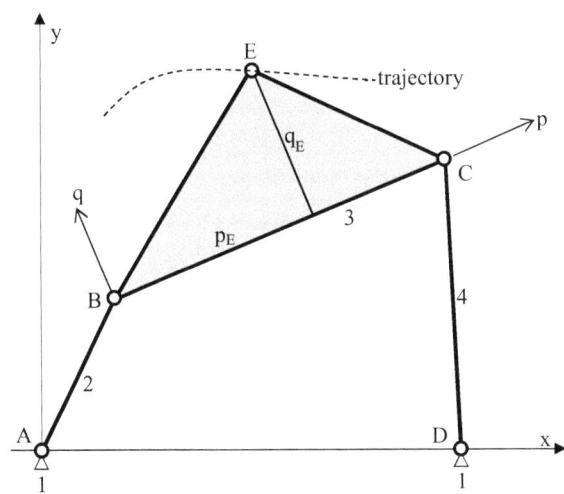

*Fig 1/2 Sketch of the four-bar mechanism*

Independent coordinates as design parameters using the markers of Fig 1/2 are the elements of the following vector ($l_2 = \overline{AB}, l_3 = \overline{BC}, l_4 = \overline{CD}$):

$$\mathbf{d}^T = [l_2, l_3, l_4, p_E, q_E] \qquad (1/1)$$

The vector of dependent coordinates is

$$\mathbf{c}^T = [x_B, y_B, x_C, y_C, x_E, y_E]. \qquad (1/2)$$

Using above parameters and coordinates geometrical and functional constraint equations can be written down. The geometrical constraint equations for point $P_i$ are

$$(x_{Bi} - x_A)^2 + (y_{Bi} - y_A)^2 - l_2^2 = 0 \qquad (1/3)$$

$$(x_{Bi} - x_{Ci})^2 + (y_{Bi} - y_{Ci})^2 - l_3^2 = 0 \qquad (1/4)$$

$$(x_{Ci} - x_D)^2 + (y_{Ci} - y_D)^2 - l_4^2 = 0 \qquad (1/5)$$

$$x_{Ei} - x_{Bi} - \frac{p_E(x_{Ci} - x_{Bi})}{l_3} + \frac{q_E(y_{Ci} - y_{Bi})}{l_3} = 0 \qquad (1/6)$$

$$y_{Ei} - y_{Bi} - \frac{p_E(y_{Ci} - y_{Bi})}{l_3} - \frac{q_E(x_{Ci} - x_{Bi})}{l_3} = 0 \qquad (1/7)$$

The functional constraint equations are created on the basis of specific requirements which should be complied. Point 3 on the coupler which draws a trajectory that passes through a series of predefined. Every single design point corresponds to different values of the elements of vector $\mathbf{c}$, for this reason vectors $\mathbf{c}_1$, $\mathbf{c}_2$, .. are different. For the $i^{th}$ position of point $E$ is

$$x_{Ei} - x_{Pi} = 0 \qquad (1/8)$$

$$y_{Ei} - y_{Pi} = 0 \qquad (1/9)$$

As it can be seen in consequence of using of natural coordinates the constraint equations are really simple. If the number of design points is less than 6, point E on the coupler of the four-bar mechanism can go exactly through the design points $P_i$.

Without going into further details in other cases the design parameters have to be computed by minimizing of the objective function. Since there is no exact solution for similar problems, for the optimal solution the least squares method can be chosen.

Using 5 predefined points the total number of constraint equation is 5x7=35 and we have also 35 unknowns (5 design parameters and elements of vector of dependent coordinates in five different positions).

With 5 design points, it is possible to construct a mechanism which exactly satisfies the requirements (functional constraints). Using the Solver in program MS Excel 2010 the following results can be obtained. The computed design parameters are:

|  | $l_2$ | $l_3$ | $l_4$ | $p_E$ | $q_E$ |
|---|---|---|---|---|---|
| Length, m | 0,4095 | 1,0811 | 1,0591 | 0,5321 | 0,5580 |

The trajectory of point $E$ applying the computed design parameters can be seen in Fig 1/3.

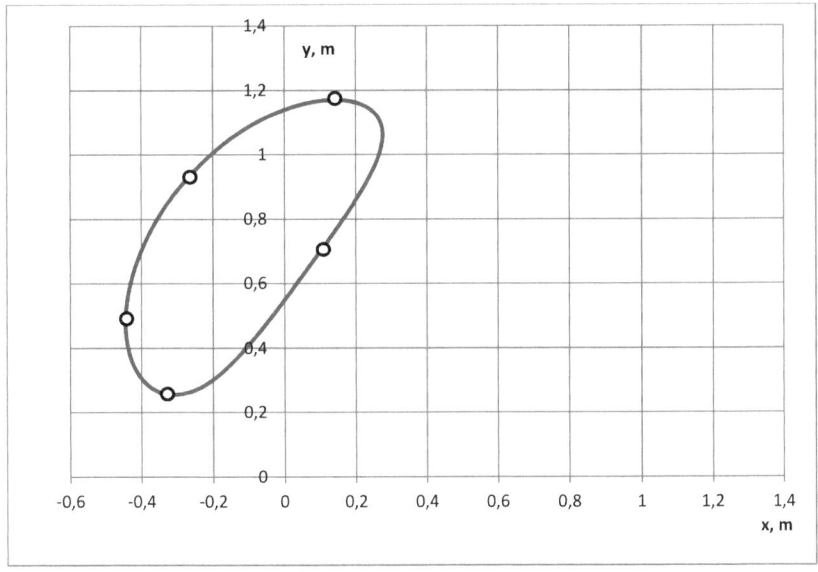

*Fig 1/3 The trajectory of point E with 5 predefined points on it*

In a general way it is needed to analyze the created structure from kinematical point of view. The driving part rotates according to kinematical function $\varphi_2 = \omega_2 t$

(uniform rotation). Using markers in Fig 1/4 the following position-time and orienta-
tion-time functions can be written down.

The loop closure equation for the sketched mechanism is

$$\mathbf{l}_2 + \mathbf{l}_3 = \mathbf{L} + \mathbf{l}_4,$$ (1/10)

respectively

$$l_2 \cos \varphi_2 + l_3 \cos \varphi_3 = L + l_4 \cos \varphi_4,$$ (1/11)

$$l_2 \sin \varphi_2 + l_3 \sin \varphi_3 = l_4 \sin \varphi_4.$$ (1/12)

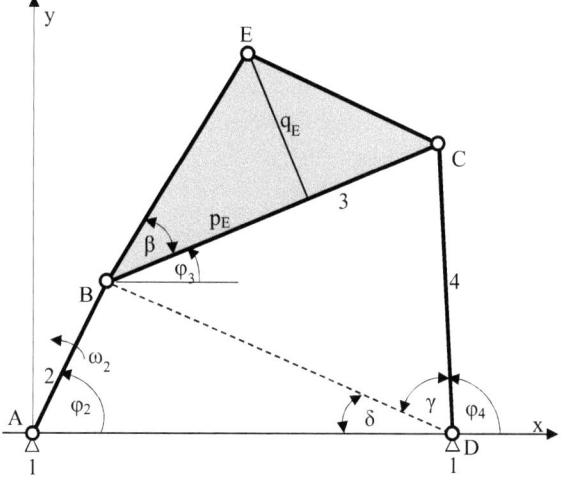

*Fig 1/4 Sketch of the created four-bar mechanism*

Furthermore

$$\beta = arctg \frac{q_E}{p_E}, \quad \sin \varphi_3 = \frac{x_C - x_B}{l_3}, \quad \cos \varphi_3 = \frac{y_C - y_B}{l_3}.$$ (1/13)

On the basis of Fig 1/4

$$L_{BD}^{\,2} = l_2^{\,2} + L^2 - 2 l_2 L \cos \varphi_2,$$ (1/14)

from which

$$\delta = arcsin\frac{l_2 \sin\varphi_2}{L_{BD}} = arcsin\frac{l_2 \sin\varphi_2}{\sqrt{l_2^2 + L^2 - 2l_2 L \cos\varphi_2}}, \quad (1/15)$$

respectively

$$l_3^2 = l_4^2 + L_{BD}^2 - 2l_4 L_{BD} \cos\gamma, \quad (1/16)$$

and

$$\gamma = arccos\frac{l_4^2 + L_{BD}^2 - l_3^2}{2L_{BD}l_4}. \quad (1/17)$$

The transmission function of the four-bar mechanism is

$$\varphi_4 = \pi - (\delta + \gamma) = \pi - (arcsin\frac{l_2 \sin\varphi_2}{L_{BD}} + arccos\frac{l_4^2 + L_{BD}^2 - l_3^2}{2L_{BD}l_4}). \quad (1/18)$$

With above quantities the elements of vector of dependent coordinates (1/2) are

$$x_B = l_2 \cos\varphi_2, \quad y_B = l_2 \sin\varphi_2, \quad (1/19)$$

$$x_C = L + l_4 \cos\varphi_4, \quad y_C = l_4 \sin\varphi_4, \quad (1/20)$$

$$x_E = l_2 \cos\varphi_2 + \sqrt{p_E^2 + q_E^2} \cos(\varphi_3 + \beta), \quad (1/21)$$

$$y_E = l_2 \sin\varphi_2 + \sqrt{p_E^2 + q_E^2} \sin(\varphi_3 + \beta). \quad (1/22)$$

All the dependent coordinates are functions of time. In some cases the transmission function of the mechanism could be interesting and/or important as well. By simple numerical differentiation to time the components of different velocities and acceleration can be obtained. MS Excel 2010 is a quite suitable solution for this task.

Cells in program MS Excel 2010 are rubricates in tables below. This simple method can be studied in the followings (time step, $\Delta t = t_{i+1} - t_i$):

## Numerical derivation of transmission function

| $t$ | $\varphi_2 = \omega_2 t$ | $\varphi_4(t)$ | $\dot{\varphi}_4(t) = \omega_4(t)$ | $\ddot{\varphi}_4(t) = \varepsilon_4(t)$ |
|---|---|---|---|---|
| $t_o$ | $\varphi_{2o}$ | $\varphi_{4o}$ | - | - |
| $t_1$ | $\varphi_{21}$ | $\varphi_{41}$ | $\dot{\varphi}_{41} = \dfrac{\varphi_{41} - \varphi_{4o}}{t_1 - t_o}$ | - |
| $t_2$ | $\varphi_{22}$ | $\varphi_{42}$ | $\dot{\varphi}_{42} = \dfrac{\varphi_{42} - \varphi_{41}}{t_2 - t_1}$ | $\ddot{\varphi}_{42} = \dfrac{\dot{\varphi}_{42} - \dot{\varphi}_{41}}{t_2 - t_1}$ |
| $t_3$ | $\varphi_{23}$ | $\varphi_{43}$ | $\dot{\varphi}_{43} = \dfrac{\varphi_{43} - \varphi_{42}}{t_3 - t_2}$ | $\ddot{\varphi}_{43} = \dfrac{\dot{\varphi}_{43} - \dot{\varphi}_{42}}{t_3 - t_2}$ |
| $t_4$ | .... | .... | .... | .... |
| $t_5$ | .... | .... | .... | .... |

## Numerical derivation of components of position function of point $E$

| $t$ | $\varphi_2 = \omega_2 t$ | $x_E(t)$ | $\dot{x}_E(t)$ | $\ddot{x}_E(t)$ | $y_E(t)$ | $\dot{y}_E(t)$ | $\ddot{y}_E(t)$ |
|---|---|---|---|---|---|---|---|
| $t_o$ | $\varphi_{2o}$ | $x_{Eo}$ | - | - | $y_{Eo}$ | - | - |
| $t_1$ | $\varphi_{21}$ | $x_{E1}$ | $\dot{x}_{E1} = \dfrac{x_{E1} - x_{Eo}}{t_1 - t_o}$ | - | $y_{E1}$ | $\dot{y}_{E1} = \dfrac{y_{E1} - y_{Eo}}{t_1 - t_o}$ | - |
| $t_2$ | $\varphi_{22}$ | $x_{E2}$ | $\dot{x}_{E2} = \dfrac{x_{E2} - x_{E1}}{t_2 - t_1}$ | $\ddot{x}_{E2} = \dfrac{\dot{x}_{E2} - \dot{x}_{E1}}{t_2 - t_1}$ | $y_{E2}$ | $\dot{y}_{E2} = \dfrac{y_{E2} - y_{E1}}{t_2 - t_1}$ | $\ddot{y}_{E2} = \dfrac{\dot{y}_{E2} - \dot{y}_{E1}}{t_2 - t_1}$ |
| $t_3$ | $\varphi_{23}$ | $x_{E3}$ | $\dot{x}_{E3} = \dfrac{x_{E3} - x_{E2}}{t_3 - t_2}$ | $\ddot{x}_{E3} = \dfrac{\dot{x}_{E3} - \dot{x}_{E2}}{t_3 - t_2}$ | $y_{E3}$ | $\dot{y}_{E3} = \dfrac{y_{E3} - y_{E2}}{t_3 - t_2}$ | $\ddot{y}_{E3} = \dfrac{\dot{y}_{E3} - \dot{y}_{E2}}{t_3 - t_2}$ |
| $t_4$ | .... | .... | .... | .... | .... | .... | .... |
| $t_5$ | .... | .... | .... | .... | .... | .... | .... |

Applying above numerical procedure the next kinematical diagrams of created mechanism can be plotted (Fig 1/5-6).

Data are the previously computed design parameters and the uniform angular velocity of driving part:

| $l_2$, m | $l_3$, m | $l_4$, m | $p_E$, m | $q_E$, m | $\omega_2$, 1/s |
|---|---|---|---|---|---|
| 0,4095 | 1,0811 | 1,0591 | 0,5321 | 0,5580 | 20 |

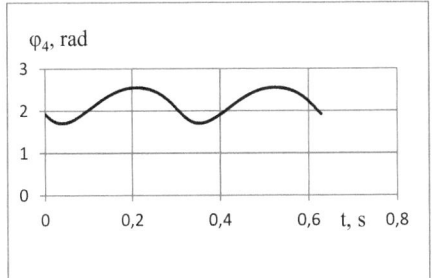

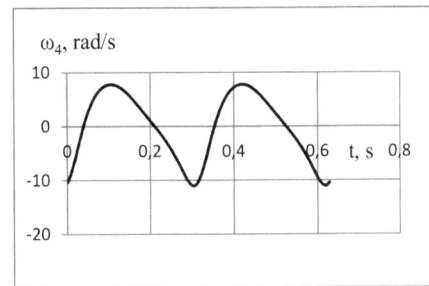

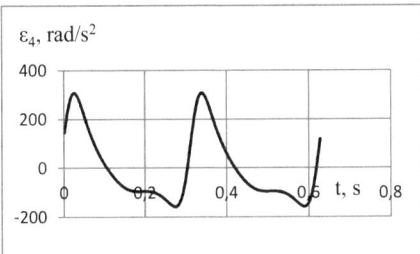

*Fig 1/5 Transmission function and its derivatives of created four-bar mechanism*

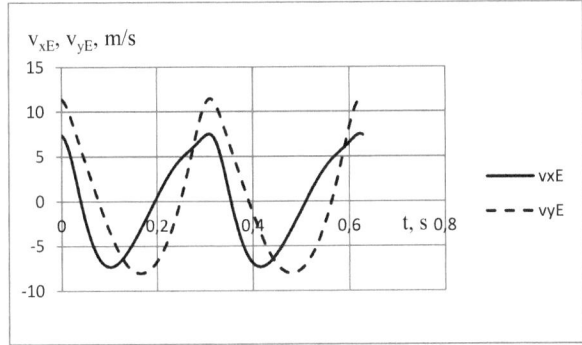

*Fig 1/6 Velocity and acceleration components of point E*

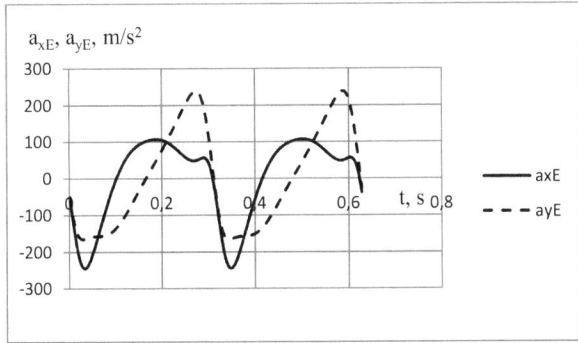

## Example 1/2 – *Kinematic synthesis of a slider-crank mechanism*

There is a sketch of slider-crank mechanism can be seen in Fig 1/7. Driving part is marked by *2*. Point *E* on the coupler should go through *5* predefined points. Their coordinates are:

|       | $P_1$ | $P_2$ | $P_3$ | $P_4$ | $P_5$ |
|-------|-------|-------|-------|-------|-------|
| x, m  | 0,8   | 1,5   | 1,8   | 1,6   | 1,0   |
| y, m  | 2,2   | 2,1   | 1,8   | 1,2   | 1,0   |

Design parameters (elements of vector **d**) have to be computed. In this example $x_A=0$, $y_A$ are design parameters.

**Important remark:** predefined points are given in coordinate system *xy* (Fig 1/7). Results depend strongly on position and orientation of applied coordinate system.

$$d^T = [\,y_A, l_2, l_3, p_E, q_E\,] \qquad (1/23)$$

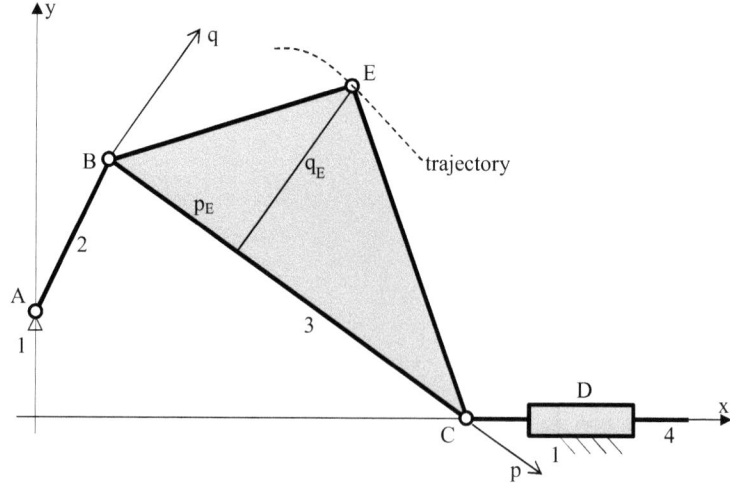

*Fig 1/7 Sketch of the slider-crank mechanism*

16

The vector of dependent coordinates is

$$\mathbf{c}^{\mathsf{T}} = [\,x_B, y_B, x_C, y_C, x_E, y_E\,]. \qquad (1/24)$$

Geometrical and functional constraint equations for point $P_i$ $(i=1, 2, \ldots\ldots, 5)$

$$(x_{Bi} - x_A)^2 + (y_{Bi} - y_A)^2 - l_2^2 = 0 \qquad (1/25)$$

$$(x_{Bi} - x_{Ci})^2 + (y_{Bi} - y_{Ci})^2 - l_3^2 = 0 \qquad (1/26)$$

$$y_{Ci} = 0 \qquad (1/27)$$

$$x_{Ei} - x_{Bi} - \frac{p_E(x_{Ci} - x_{Bi})}{l_3} + \frac{q_E(y_{Ci} - y_{Bi})}{l_3} = 0 \qquad (1/28)$$

$$y_{Ei} - y_{Bi} - \frac{p_E(y_{Ci} - y_{Bi})}{l_3} - \frac{q_E(x_{Ci} - x_{Bi})}{l_3} = 0 \qquad (1/29)$$

$$x_{Ei} - x_{Pi} = 0 \qquad (1/30)$$

$$y_{Ei} - y_{Pi} = 0 \qquad (1/31)$$

The number of constraint equations and unknowns in this example also is *35-35*. Computed design parameters:

| | $y_A$ | $l_2$ | $l_3$ | $p_E$ | $q_E$ |
|---|---|---|---|---|---|
| Length, m | 1,5695 | 0,8231 | 4,2125 | 0,9218 | 0,4235 |

Trajectory of point *E* with above design parameters can be seen in Fig 1/8. The kinematical investigation of the created slider-crank mechanism could be important.

The crank rotates according to kinematical function $\varphi_2 = \omega_2 t$ (uniform rotation). With markers in Fig 1/9 for some dependent coordinates the following position-time are obtained.

17

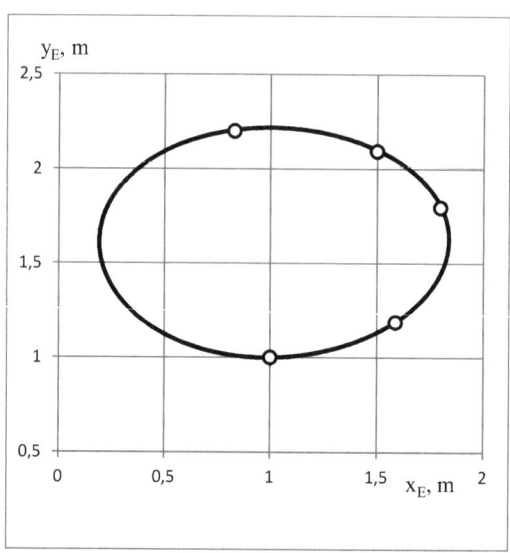

*Fig 1/8 The trajectory of point E with 5 predefined points on it (slider-crank mechanism)*

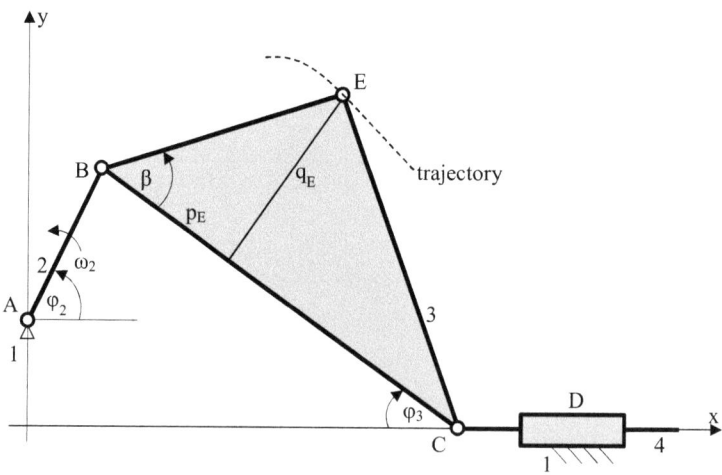

*Fig 1/9 Sketch of the created slider-crank mechanism to the kinematical investigation*

18

On the basis of Fig 1/9

$$\varphi_3 = arcsin\frac{y_A + l_2\,sin\varphi_2}{l_3}. \qquad (1/32)$$

The transmission function and other elements of vector of dependent coordinates of the slider-crank mechanism are

$$x_B = l_2\,cos\varphi_2, \quad y_B = y_A + l_2\,sin\varphi_2, \qquad (1/33)$$

$$x_C = l_2\,cos\varphi_2 + \sqrt{l_3^2 - (y_A + l_2\,sin\varphi_2)^2}, \quad y_C = 0, \qquad (1/34)$$

$$x_E = l_2\,cos\varphi_2 + \sqrt{p_E^2 + q_E^2}\,cos(\beta - \varphi_3), \qquad (1/35)$$

$$y_E = y_A + l_2\,sin\varphi_2 + \sqrt{p_E^2 + q_E^2}\,sin(\beta - \varphi_3). \qquad (1/36)$$

By simple numerical differentiation respecting to time the components of different velocities and acceleration can be obtained again.

Applying the numerical procedure presented in previous example the next kinematical diagrams can be plotted (Fig 1/10-13).

Data are the computed design parameters and the uniform angular velocity of rotating crank:

| $y_A$, m | $l_2$, m | $l_3$, m | $p_E$, m | $q_E$, m | $\omega_2$, 1/s |
|---|---|---|---|---|---|
| 1,5695 | 0,8231 | 4,2125 | 0,9218 | 0,4235 | 20 |

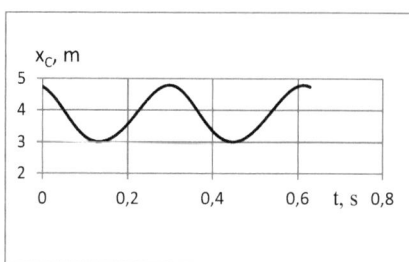

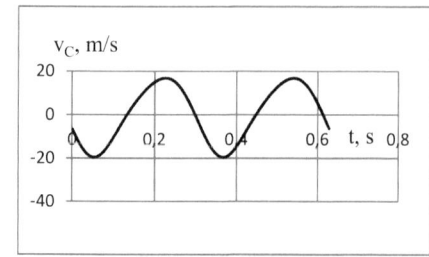

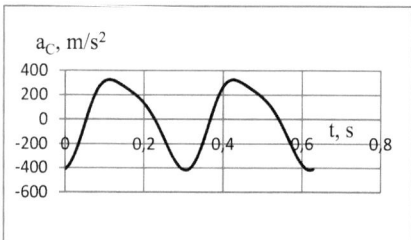

*Fig 1/10 Transmission function and its derivatives of created slider-crank mechanism*

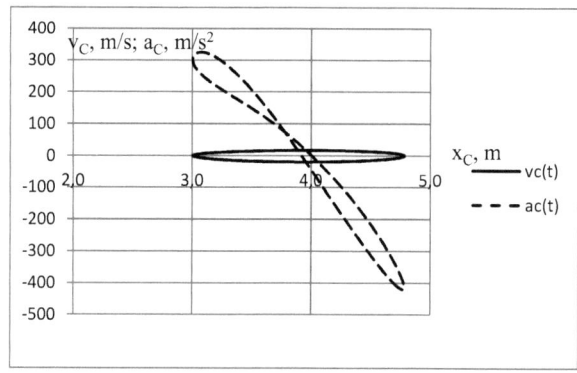

*Fig 1/11 Velocity and acceleration in function of position of point C on coupler of created slider-crank mechanism*

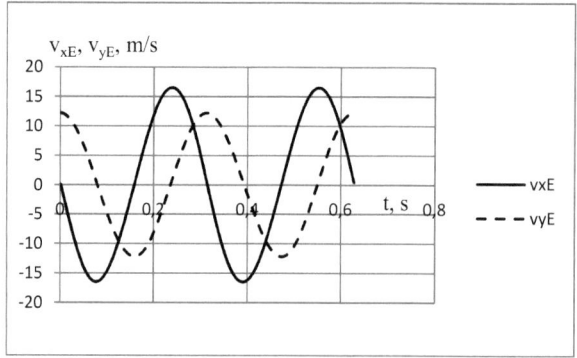

*Fig 1/12 Velocity components of point E*

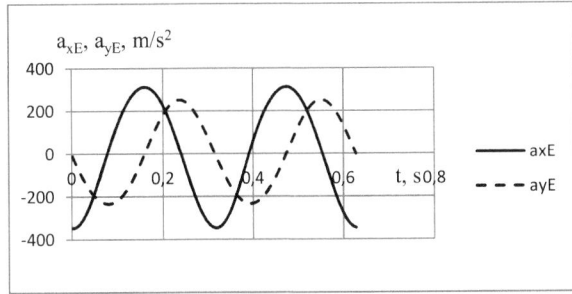

*Fig 1/13 Acceleration components of point E*

## 1.2. Function generation synthesis

This kind of synthesis is based on transmission functions. During the synthesis of planar four-bar linkages it is needed to determine the design parameters (Fig 1/14) as distances ($l_1$, $l_2$, $l_3$, $l_4$) between successive axes of applied hinges. As a result of synthesis we can get to an approximation of an expected function between input angle $\varphi_2$ and output angle $\varphi_4$.

The connection between the input and output sides can be expressed by a function of $y=f(x)$. Under limited conditions the four-bar mechanism is able to generate functions mechanically like $y=x^n$, $y=log_{10}x$, $y=e^x$ and so on. The intervals (from the initial to the final position) for the variables are $x_i \leq x \leq x_f$, $y_i \leq y \leq y_f$ and input and output motion, $\varphi_{2i} \leq \varphi_2 \leq \varphi_{2f}$, $\varphi_{4i} \leq \varphi_4 \leq \varphi_{4f}$.

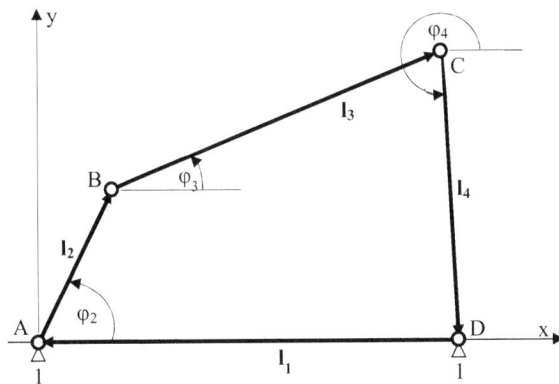

*Fig 1/14 Loop closure diagram of a four-bar linkage ($\varphi_1=180^o$)*

Before examples it is necessary to present the Freudenstein's Equation. On the basis of Fig 1/14

$$\mathbf{l}_1 + \mathbf{l}_2 + \mathbf{l}_3 + \mathbf{l}_4 = 0. \tag{1/37}$$

This is the loop closure equation which is the sum of position vectors around the structure. The vector equation can be replaced by two scalar equations such as

$$-l_1 + l_2\cos\varphi_2 + l_3\cos\varphi_3 + l_4\cos\varphi_4 = 0, \tag{1/38}$$

$$l_2\sin\varphi_2 + l_3\sin\varphi_3 + l_4\sin\varphi_4 = 0. \tag{1/39}$$

After re-arrangement

$$l_3\cos\varphi_3 = l_1 - l_2\cos\varphi_2 - l_4\cos\varphi_4, \tag{1/40}$$

$$l_3\sin\varphi_3 = -l_2\sin\varphi_2 - l_4\sin\varphi_4. \tag{1/41}$$

Squaring and adding the equations and after further re-arrangement angle $\varphi_3$ can be eliminated. Finally the following equation (Freudenstein's Equation) can be obtained:

$$\frac{l_3^2 - l_1^2 - l_2^2 - l_4^2}{2l_2 l_4} + \frac{l_1}{l_4}\cos\varphi_2 + \frac{l_1}{l_2}\cos\varphi_4 - \cos(\varphi_2 - \varphi_4) = 0. \tag{1/42}$$

There are all of design parameters in the equation. To create the desired linkage it is enough to calculate three of them. Magnitude of fourth one – for example $l_1$ – could be $1m$. The angles in the structure are independent from scaling up or down.

Applying the equation we can have three accuracy points by corresponding values of $\varphi_2$ and $\varphi_4$. In this way three equations can be written down to calculate three unknowns ($l_2$, $l_3$, $l_4$).

**Example 1/3** – Design a four-bar linkage to generate the function

| $y = x^{0.7}$ over the range $1 \leq x \leq 2$. | (1/43) |
|---|---|

Initial and final joint angles are $\varphi_{2i} = 50^\circ \leq \varphi_2 \leq \varphi_{2f} = 200^\circ$, $\varphi_{4i} = 240^\circ \leq \varphi_4 \leq \varphi_{4f} = 310^\circ$ (Fig 1/14).

At first it is necessary to select the three accuracy points. By a good choice the difference between the mathematical and transmission function can be reduced. From this point of view the Chebyshev Spacing is good choice. It is an equal spacing around a circle then projection onto the bisector of the circle (Fig 1/15).

If a range of $x_i \leq x \leq x_f$ is to be covered then $n$ accuracy points are given by

$$x_j = \frac{1}{2}(x_f + x_i) - \frac{1}{2}(x_f - x_i)\cos\frac{(2j-1)\pi}{2n}, \quad j = 1, 2, \dots n \,. \qquad (1/44)$$

In this example $n=3$ and $x_i=1$, $x_f=2$, $y_i=1$, $y_f=1,6245$. Using above equation the accuracy points are $x_1=1,06699$, $x_2=1,5$, $x_3=1,9330$, respectively $y_1=1,0464$, $y_2=1,3282$, $y_3=1,5862$.

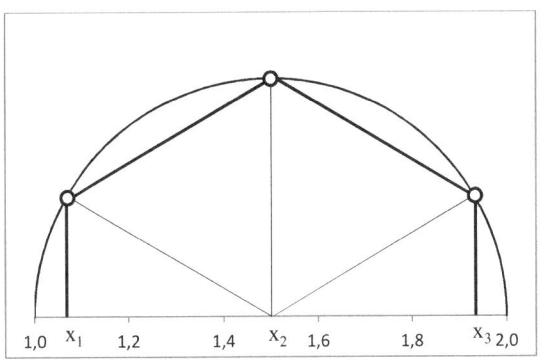

*Fig 1/15 Three accuracy points with Chebyshev spacing over the range $1 \leq x \leq 2$. In the Fig the half a regular hexagon can be seen*

As second step it is needed to transform the function $y=f(x)$ into angles $\varphi_2$ and $\varphi_4$ using the method of linear mapping. The form of equations of linear mapping is the following:

| $\varphi_2 = m_1 x + c_1, \qquad \varphi_4 = m_2 y + c_2$ . | (1/45) |
|---|---|

With initial and final values of angles $\varphi_2$ and $\varphi_4$ four linear equations can be written down:

| | | |
|---|---|---|
| $\varphi_{2i} = m_1 x_i + c_1,$ | $\varphi_{2f} = m_1 x_f + c_1,$ | (1/46) |
| $\varphi_{4i} = m_2 y_i + c_2,$ | $\varphi_{4f} = m_2 y_f + c_2,$ | (1/47) |

or with numbers of this example,

| | | |
|---|---|---|
| $50 = m_1 + c_1,$ | $200 = 2m_1 + c_1,$ | (1/48) |
| $240 = m_2 + c_2,$ | $310 = 1,6245m_2 + c_2.$ | (1/49) |

Roots of the simple equation-system: scaling parameters, $m_1=150$, $m_2=112,09$ and bias parameters, $c_1=-100$, $c_2=127,91$. With them the equations of linear mapping,

| | |
|---|---|
| $\varphi_{2j} = 150x_j - 100,$  $\varphi_{4j} = 112,09y_j + 127,91,$  $j = 1,2,3.$ | (1/50) |

Applying the equations (1/50) of linear mapping three function accuracy points to corresponding accuracy joint angles of $\varphi_2$ and $\varphi_4$ can be calculated. Details can be seen in table below.

| Position | $x$ | $\varphi_2 [^o]$ | $y$ | $\varphi_4 [^o]$ |
|---|---|---|---|---|
| initial | 1 | 50 | 1 | 240 |
| j=1 | 1,06699 | 60,05 | 1,0464 | 245,2 |
| j=2 | 1,5 | 125 | 1,3282 | 276,74 |
| j=3 | 1,93301 | 189,95 | 1,5862 | 305,65 |
| final | 2 | 200 | 1,6245 | 310 |

Table 1 Function values and the corresponding accuracy joint angles

Finally using values of joint angles of $\varphi_2$ and $\varphi_4$ in the three accuracy points three Freudenstein's Equations are at our disposal.

| | |
|---|---|
| $\dfrac{l_3^2 - l_1^2 - l_2^2 - l_4^2}{2l_2l_4} + \dfrac{l_1}{l_4}\cos\varphi_{2j} + \dfrac{l_1}{l_2}\cos\varphi_{4j} - \cos(\varphi_{2j} - \varphi_{4j}) = 0,$  $j = 1, 2, 3.$ | (1/51) |

If $l_1=1$ m, the remaining unknowns (design parameters) can be determined – for example – by the aid of Solver in MS Excel. Computed results are the followings:

| Link | $l_1$ | $l_2$ | $l_3$ | $l_4$ |
|--------|-------|--------|--------|--------|
| length, m | 1 | 0,8231 | 4,2125 | 0,9218 |

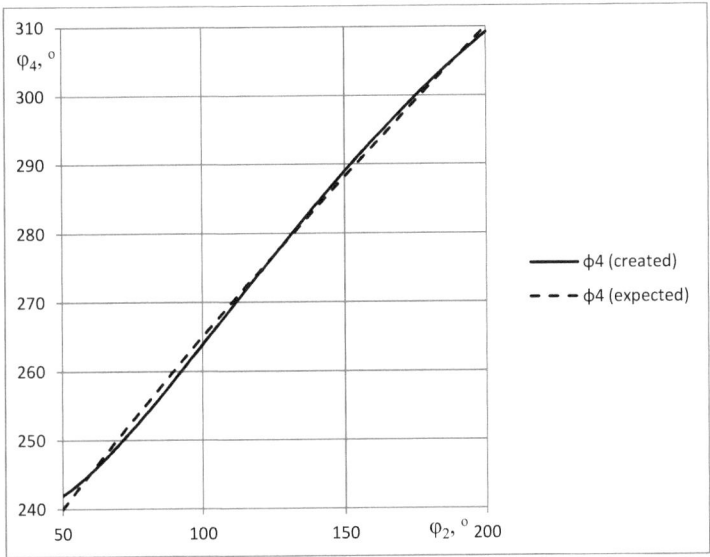

*Fig 1/16 Connection between joint angles of $\varphi_2$ and $\varphi_4$ ($y=x^{0,7}$)*

As it was mentioned above the lengths of links can be scaled up or down. Sometimes the results of first effort are not acceptable. For example some links could be incommensurate and/or some other links or Grashof-condition for a four-bar linkage is not satisfied.

In these cases the position and magnitude of interval of joint angles of $\varphi_2$ and $\varphi_4$ have to be modified and try it again. Reasonable results can be obtained after some trying.

Important remark: from point of view of successful design the good selection of intervals of joint angles (Fig 1/14) is necessary. Some experience of designer could help.

In Fig 1/16 the connection between joint angles of $\varphi_2$ and $\varphi_4$ can be seen. The transmission function of created linkages is illustrated by continuous curve and the expected one by dashed curve. The difference between them cannot be mentioned significant. It is easy to notice the three accuracy points.

**Example 1/4** – Design a four-bar linkage to generate the function

| y=log₁₀x over the range $1 \leq x \leq 2$. | (1/52) |
|---|---|

Initial and final joint angles are $\varphi_{2i}- 60^\circ \leq \varphi_2 \leq \varphi_{2f}= 120^\circ$, $\varphi_{4i}= 100^\circ \leq \varphi_4 \leq \varphi_{4f}= 70^\circ$ (Fig 1/17).

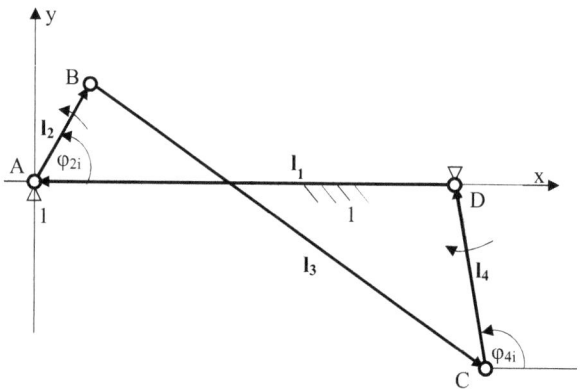

*Fig 1/17 Sketch of a four-bar linkage in initial position with the loop closure diagram*
*($\varphi_1=180^\circ$)*

The selection of the three accuracy points has to be the first step: $n=3$ and $x_i=1$, $x_f=2$, $y_i=logx_i=log1=0$, $y_f= logx_f=log2=0,301$. Using equation the accuracy points (1/44) are $x_1=1,06699$, $x_2=1,5$, $x_3=1,9330$, respectively $y_1=0,0282$, $y_2=0,1761$, $y_3=0,2862$.

The next step is the linear mapping between the function $y=f(x)$ into angles $\varphi_2$ and $\varphi_4$. On the basis of equations of linear mapping (1/45) applying the initial and final values of angles $\varphi_2$ and $\varphi_4$ the linear equation-system is

| $60 = m_1 +c_1,$     $120 = 2m_1 +c_1$ . | (1/53) |
|---|---|
| $100 = c_2,$     $70 = 0,301m_2 +c_2$ . | (1/54) |

Roots of the equation-system: scaling parameters, $m_1=60$, $m_2=-99,67$ and bias parameters, $c_1=0$, $c_2=100$ and with them the equations of linear mapping,

| $\varphi_{2j} = 60x_j,$     $\varphi_{4j} = -99,67y_j +100,$   $j =1,2,3$ . | (1/55) |
|---|---|

With equations (1/55) of linear mapping the three function accuracy points to corresponding accuracy joint angles of $\varphi_2$ and $\varphi_4$ can be calculated (Table 2).

| Position | x | $\varphi_2 [^o]$ | y | $\varphi_4 [^o]$ |
|----------|---|---------|---|---------|
| initial | 1 | 60 | 0 | 100 |
| j=1 | 1,06699 | 64,02 | 0,02816 | 97,19 |
| j=2 | 1,5 | 90 | 0,1761 | 82,45 |
| j=3 | 1,93301 | 115,98 | 0,2862 | 71,47 |
| final | 2 | 120 | 0,3010 | 70 |

Table 2 Function values and the corresponding accuracy joint angles ($y=log_{10}x$)

Finally three Freudenstein's Equations (1/51) with values of joint angles of $\varphi_2$ and $\varphi_4$ in the three accuracy points can be written down. Supposing that $l_1=1$ m, computed results are listed below:

| Link | $l_1$ | $l_2$ | $l_3$ | $l_4$ |
|------|-------|-------|-------|-------|
| length, m | 1 | 0,16218 | 1,06700 | 0,30685 |

In Fig 1/18 the connection between joint angles of $\varphi_2$ and $\varphi_4$ can be seen concerning the transmission function of created linkages and the expected curve.

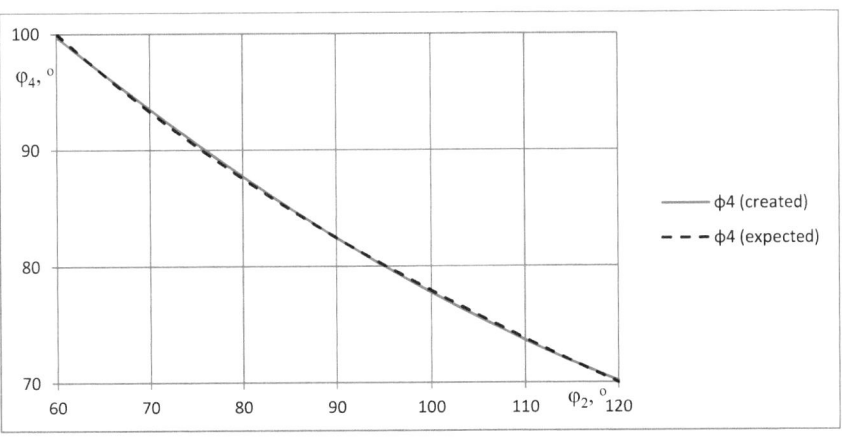

Fig 1/18 Connection between joint angles of $\varphi_2$ and $\varphi_4$ ($y=logx$)

**Example 1/5** – Design an offset slider-crank mechanism to generate the function

$$y=x^{1.5} \text{ over the range } 1 \leq x \leq 2. \tag{1/56}$$

Initial and final joint angles of the crank are $\varphi_i = 45^\circ \leq \varphi \leq \varphi_f = 105^\circ$, further positions of slider are $s_i = 1,000 \ m \leq s \leq s_f = 0,700 \ m$ (Fig 1/19).

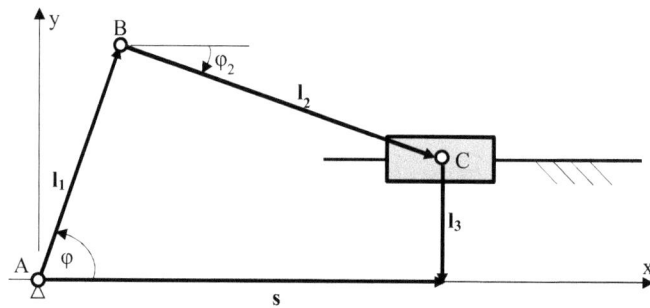

*Fig 1/19 Sketch of an offset slider-crank mechanism with the loop closure diagram*

The loop closure equation on the basis of sketch of offset slider-crank mechanism can be seen in Fig 1/19 is

$$\mathbf{l_1 + l_2 + l_3 = s}, \tag{1/57}$$

which can be replaced by two scalar equations such as

$$l_1 \cos\varphi + l_2 \cos\varphi_2 = s, \tag{1/58}$$

$$l_1 \sin\varphi + l_2 \sin\varphi_2 - l_3 = 0. \tag{1/59}$$

After re-arrangement

$$l_2 \cos\varphi_2 = s - l_1 \cos\varphi, \quad l_2 \sin\varphi_2 = l_3 - l_1 \sin\varphi. \tag{1/60}$$

In order to eliminate angle $\varphi_2$ is expedient to square and add the equations (1/60). Finally the following equation can be written down:

28

$$s^2 + l_1^2 + l_3^2 - l_2^2 - 2l_1l_3\sin\varphi - 2sl_1\cos\varphi = 0. \qquad (1/61)$$

In equation (1/61) all of design parameters ($l_1$, $l_2$, $l_3$) can be found. Using the equation and the three accuracy points with corresponding values of $\varphi_2$ and $\varphi_4$ three equations can be written down to calculate the three unknowns.

After selection of the three accuracy points (Chebyshev spacing) $x_i=1$, $x_f=2$, $y_i=x_i^{1,5}=1^{1,5}=1$, $y_f=x_f^{1,5}=2^{1,5}=2,8284$. The accuracy points on the basis of equation (1/44) are $x_1=1,06699$, $x_2=1,5$, $x_3=1,9330$, respectively $y_1=1,10215$, $y_2=1,83712$, $y_3=2,68750$.

The linear mapping is needed between the function $y=f(x)$ into angle of crank $\varphi_2$ and position of slider $s$. According to equations of linear mapping (1/45) and the initial and final values $\varphi$ and $s$ the linear equation-system is

$$45 = m_1 + c_1, \qquad 105 = 2m_1 + c_1, \qquad (1/62)$$

$$1 = m_2 + c_2, \qquad 0,7 = 2,8284m_2 + c_2. \qquad (1/63)$$

Roots as scaling parameters are $m_1=60$, $m_2=-0,16408$ and bias parameters, $c_1=-15$, $c_2=1,6408$. With them the equations of linear mapping are

$$\varphi_j = 60x_j - 15, \qquad s_j = -0,16408y_j + 1,16408 \qquad j = 1,2,3. \qquad (1/64)$$

Applying equations (1/64) of linear mapping the three function accuracy points to corresponding accuracy joint angle $\varphi$ and position of piston $s$ can be calculated (Table 3).

| Position | x | $\varphi$ [°] | y | s [m] |
|---|---|---|---|---|
| initial | 1 | 45 | 1 | 1,0000 |
| j=1 | 1,06699 | 49,02 | 1,10215 | 0,9832 |
| j=2 | 1,5 | 75 | 1,83712 | 0,8624 |
| j=3 | 1,93301 | 100,98 | 2,68750 | 0,7231 |
| final | 2 | 105 | 2,8284 | 0,7000 |

Table 3 Function values and the corresponding accuracy joint angle and slider position ($y=x^{1,5}$)

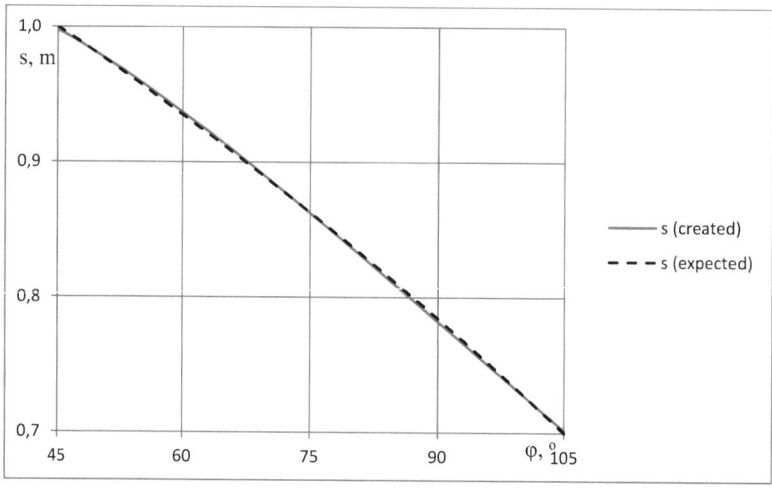

*Fig 1/20 Connection between joint angle of $\varphi$ and position of slider of s $(y=x^{1,5})$*

Written down equation (1/51) three times with values of joint angle of $\varphi$ and position of slider $s$ in the three accuracy points we can get to the desired design parameters. Computed results are listed here:

| Link | $l_1$ | $l_2$ | $l_3$ |
|---|---|---|---|
| length, m | 0,30993 | 0,78221 | 0,29337 |

In Fig 1/20 the connection between joint angle of $\varphi$ and slider displacement position $s$ can be seen concerning the transmission function of created structure and the expected curve.

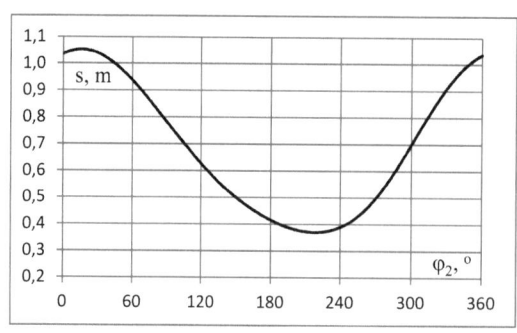

*Fig 1/21 Slider displacement position (created mechanism)*

The slider displacement position in function of angle position of crank can be seen in Fig 1/21. The Grashof-condition is satisfied apparently ($l_3 < l_2 - l_1$).

30

## 2. Some dynamic examples in mechanical engineering

### 2.1. Nonlinear mechanical vibration

### 2.1.1. Nonlinear mechanical free vibration

In Fig 2/1 the simplest mass-spring model consisting of a body (mass *m*) and a nonlinear spring can be seen. The movement of the body is influenced only by spring force $\mathbf{F}_{sp}$. The friction between body and horizontal surface furthermore the own mass of the spring are neglected.

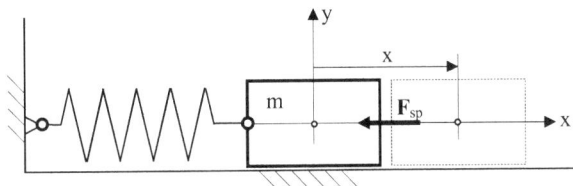

*Fig 2/1 Spring-mass model (plotted displacement and length of the spring are not proportional)*

Starting from the sketched elongated position the body makes free oscillating motion on constant natural frequency. It is possible only for conservative systems or with other words: there is no transfer of energy or out feed.

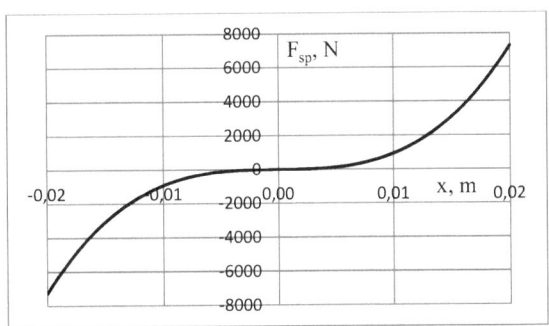

*Fig 2/2 Example for characteristics of strongly nonlinear spring*

In unloaded position the origin of coordinate-system *xy* fastened to the base and the center of mass of the body are coincided. For lengthening of the spring force is needed. The connection between force and lengthening is demonstrated by characteristic curve of spring (Fig 2/2). The characteristics of the nonlinear spring can be described by equation

$$F_{sp} = ax + bx^3 .$$ (2/1)

As it can be seen the spring is strongly nonlinear. In equation (2/1): $a=3000$ N/m, $b=300\,000$ N/m$^3$.

In Fig 2/1 the oscillating body can be seen in displaced position. The body moves on the effect of spring force which direction is opposite to the displacement. According to basic law of dynamics

$$- F_{sp} = -( ax + bx^3 ) = ma = m\ddot{x} ,$$ (2/2)

from which

$$m\ddot{x} + ax + bx^3 = 0 .$$ (2/3)

This is the second order differential-equation of nonlinear free vibration concerning the characteristics of spring described in equation (2/1). Its analytical solution in most cases is quite difficult or impossible. The simplest choice to solve it is applying numerical methods. In such cases the application of numerical methods is advantageous. The results obtained in this way can be demonstrated in different kinematical diagrams.

Oscillating system can be described by translational coordinate $x$. To the solution it is needed to determine the initial conditions $x_o$, $\dot{x}_o$. Time step: $t_{i+1}-t_i$. Applied algorithms in MS Excel can be seen in table below.

| $t$ | $\ddot{x}(x)$ | $\dot{x}$ | $x$ |
|---|---|---|---|
| $t_o$ | $\ddot{x}_o(x_o)$ | $\dot{x}_o$ | $x_o$ |
| $t_1$ | $\ddot{x}_1(x_1)$ | $\dot{x}_1 = \dot{x}_o + \ddot{x}_o(t_1 - t_o)$ | $x_1 = x_o + \dot{x}_1(t_1 - t_o)$ |
| $t_2$ | $\ddot{x}_2(x_2)$ | $\dot{x}_2 = \dot{x}_1 + \ddot{x}_1(t_2 - t_1)$ | $x_2 = x_1 + \dot{x}_2(t_2 - t_1)$ |
| $t_3$ | .... | .... | .... |

**Example 2/1** – Using above numerical algorithms plot kinematical functions of oscillating body can be shown in Fig 2/1. *Data: m=8 kg, a=3000 N/m, b=300 000 N/m³;* initial conditions: $x_o = 0,020m$, $\dot{x}_o = 0\,m/s$.

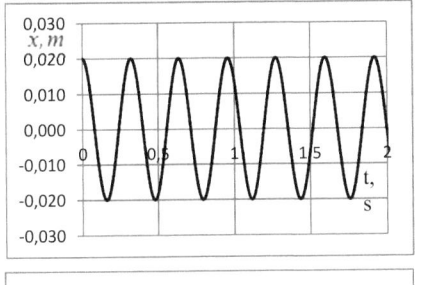

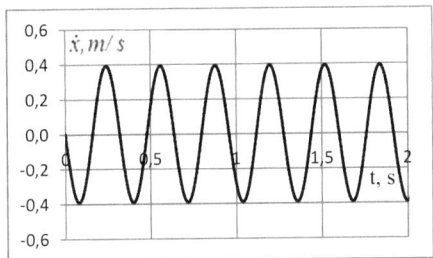

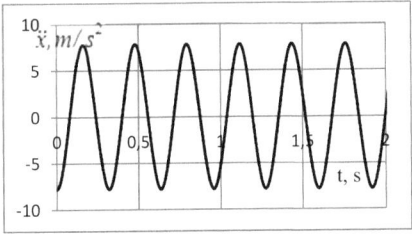

Fig 2/3 Kinematical diagrams of oscillating body (0 s ≤ t ≤ 2 s)

Seeing diagrams in Fig 2/3 it can be noticed that the translational motion is periodic but not harmonic motion due to the strongly nonlinear spring.

## 2.1.2. Nonlinear damped free vibration

Until now it was supposed that the energy level of the oscillating system is constant. The body moves on the effect of spring force. In the reality there are some kind of resistance between the body and its environment: air resistance, frictional resistance and dampers.

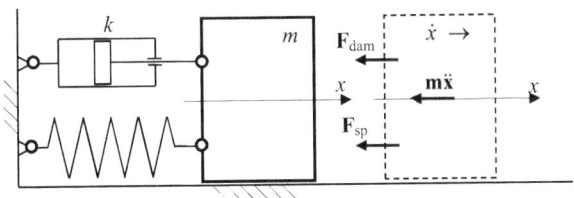

Fig 2/4 Sketch of viscous damped nonlinear oscillating system

In this textbook it is about viscous dampers. Dampers decrease continuously the energy level of oscillating system for this reason decreases the amplitude of oscillation and finally it dies away.

Damping is caused by damping force $F_{dam}$ (Fig 2/4). In case of most frequently applied viscous damping the damping force is proportional to velocity of oscillating body and its direction is opposite to the velocity of the body, i. e. $F_{dam} = -k\dot{x}$ where $k$ is damping factor. Its dimensional unit: $Ns/m$.

Motion equation of oscillating body is

$$m\ddot{x} + k\dot{x} + ax + bx^3 = 0. \qquad (2/4)$$

Above numerical algorithms can be applied again. In table below the motion equation differs from the previous one.

| $t$ | $\ddot{x}(x,\dot{x})$ | $\dot{x}$ | $x$ |
|---|---|---|---|
| $t_o$ | $\ddot{x}_o(x_o,\dot{x}_o)$ | $\dot{x}_o$ | $x_o$ |
| $t_1$ | $\ddot{x}_1(x_1,\dot{x}_1)$ | $\dot{x}_1 = \dot{x}_o + \ddot{x}_o(t_1 - t_o)$ | $x_1 = x_o + \dot{x}_1(t_1 - t_o)$ |
| $t_2$ | $\ddot{x}_2(x_2,\dot{x}_2)$ | $\dot{x}_2 = \dot{x}_1 + \ddot{x}_1(t_2 - t_1)$ | $x_2 = x_1 + \dot{x}_2(t_2 - t_1)$ |
| $t_3$ | .... | .... | .... |

**Example 2/2** – Plot kinematical diagrams of viscous damped free oscillating motion (Fig 2/4) applying above numerical method. Data: $m=4\ kg$, $k=10\ Ns/m$, $a=5000\ N/m$, $b=3000\ 000\ N/m^3$; initial conditions: $x_o = 0\ m$, $\dot{x}_o = 6\ m/s$.

Plotted kinematical diagrams can be studied in Fig 2/5. Some remarks to them:

➤ The amplitude of oscillation decreases and finally it dies away.
➤ By reduction of amplitude the frequency of the oscillation is getting lower and lower due to nonlinear progressive spring.
➤ The oscillation is periodical but strongly not harmonic.

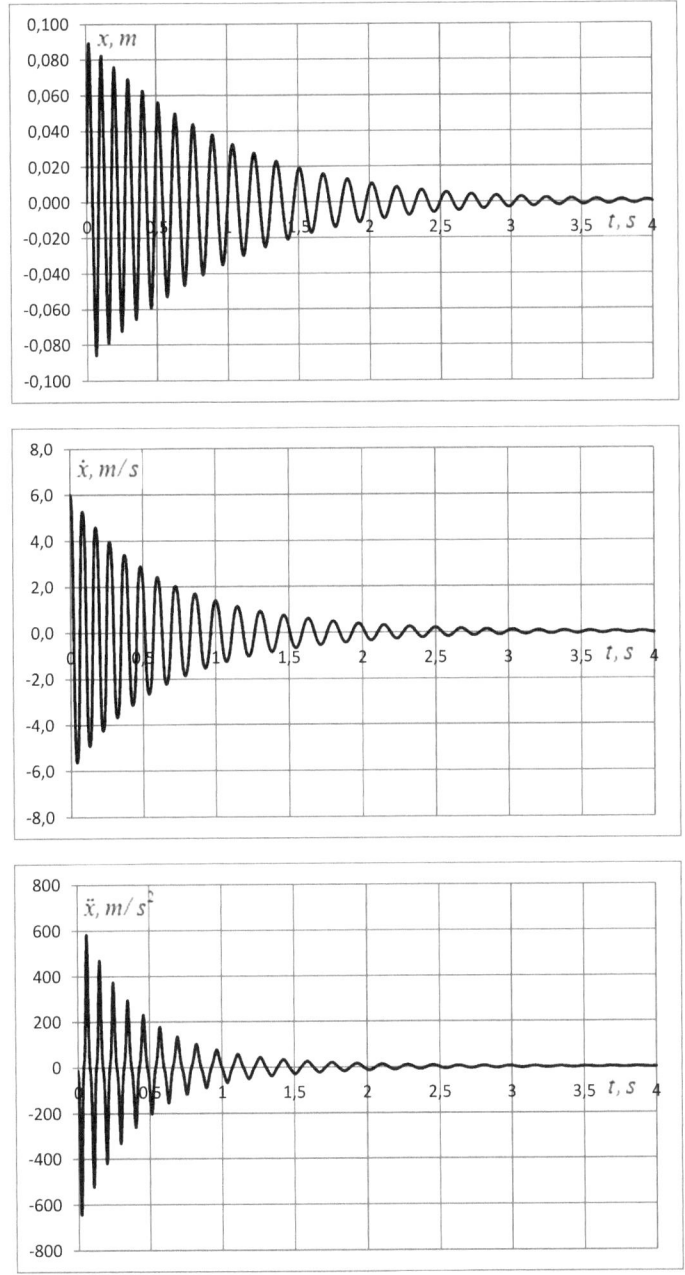

*Fig 2/5 Kinematical diagrams of damped free oscillation (0 s ≤ t ≤ 4 s)*

### 2.1.3. Comparison of numerical and analytical solution – in linear case

In order to check up on the accuracy of applied numerical method let us see Fig 2/4 again. But now be the spring a linear one; its spring stiffness is denoted by $s$. The motion equation of oscillating body is

$$m\ddot{x} + k\dot{x} + sx = 0. \tag{2/5}$$

If the initial conditions are $x_o = A_o$, $\dot{x}_o = 0$ the analytical solution of equation (2/5) is

$$x(t) = Ae^{-\beta t} \cos \delta t. \tag{2/6}$$

In the equation above $k$ is damping factor, $\beta = \dfrac{k}{2m}$ is damping constant, $\delta = \sqrt{\omega^2 - \beta^2}$ is the angle frequency of damped oscillation and $\omega$ is the natural angle frequency of free oscillation.

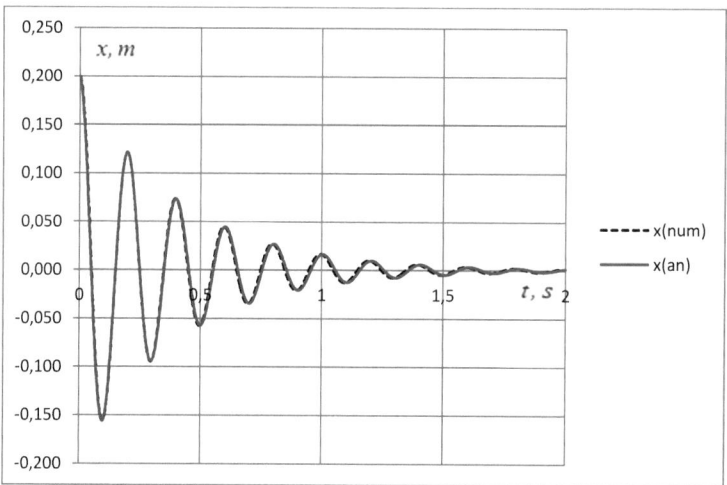

*Fig 2/6 x-t diagrams for viscous damped oscillator obtained analytically and numeri-cally (0 s ≤ t ≤ 2 s), applied spring has linear characteristics*

**Example 2/3** – In order to compare numerical method applied in this chapter and an-alytical solution according to equation (2/6) plot position-time diagram of viscous

36

damped free oscillating motion (Fig 2/4). The spring has linear characteristics. Data: $m=4\ kg,\ k=20\ Ns/m,\ s=4000\ N/m$, initial conditions: $x_o =0{,}2\ m,\ \dot{x}_o =0\ m/s$.

In Fig 2/6 the x-t diagrams obtained analytically and numerically are plotted. It can be seen that the difference is negligible.

## 2.1.4 Example for geometrical nonlinearity

In Fig 2/7 a mass-spring model can be seen. In this case the spring is linear and it is located in slanting position described by angle $\varphi$. The mass point can move along the vertical trajectory. Its movement is influenced by gravitational force and vertical component of spring force. The friction between mass point and vertical surface furthermore the own mass of the spring are neglected.

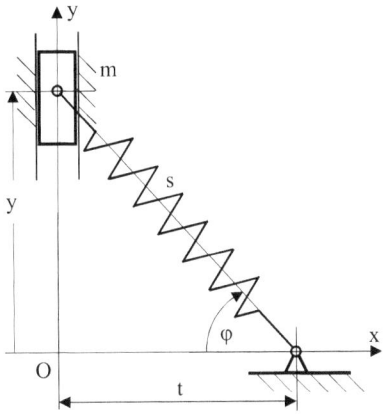

*Fig 2/7 Sketch of oscillating system*

The position of mass point can be described by coordinate $y$. On the basis of above mentioned its motion equation is

$$-mg+F_{sp}\sin\varphi = m\ddot{y}, \qquad (2/7)$$

where spring force is

$$F_{sp} = s(l_o -l)= s(l_o -\sqrt{y^2 +t^2}\,), \qquad (2/8)$$

where $l$ the actual, $l_o$ unloaded length of the spring. Considering that

$sin\ \varphi = \dfrac{y}{\sqrt{y^2+t^2}}$ , finally the

$$m\ddot{y} = s(l_o - \sqrt{y^2+t^2})\dfrac{y}{\sqrt{y^2+t^2}} - mg , \qquad (2/9)$$

motion equation can be obtained from which

$$\ddot{y} = \dfrac{s}{m}(\dfrac{l_o}{\sqrt{y^2+t^2}} - 1)y - g . \qquad (2/10)$$

**Example 2/4** – Plot the kinematical diagrams of oscillating mass point in case of small and big displacements (Fig 2/7). The spring has linear characteristics in this case as well. Data: $m=10\ kg,\ s=4000\ N/m\ l_o=0,5\ m,\ g=9,81\ m/s^2,\ t=0,3\ m.$

a) initial conditions **in case of small displacement**: $y_o = 0,4\ m,\ \dot{y}_o = 0\ m/s$

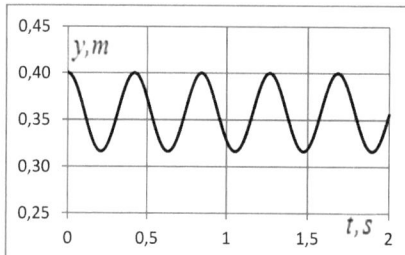

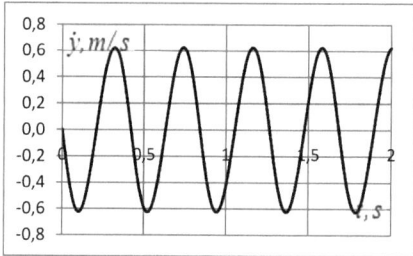

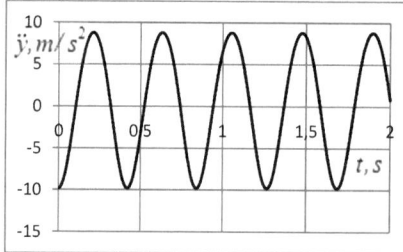

*Fig 2/8 Kinematical diagrams of oscillating mass point in case of small displacement ($0\ s \le t \le 2\ s$)*

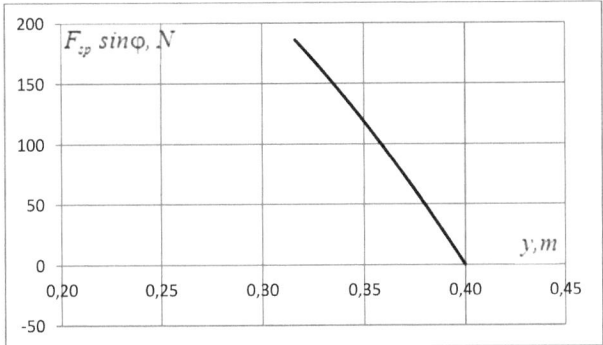

*Fig 2/9 Vertical component of spring force in function of position of mass point (small displacements)*

b) initial conditions **in case of big displacement**: $y_o = 0{,}6\,m$, $\dot{y}_o = 0\,m/s$

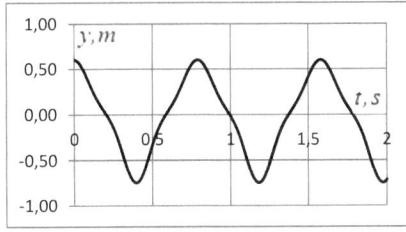

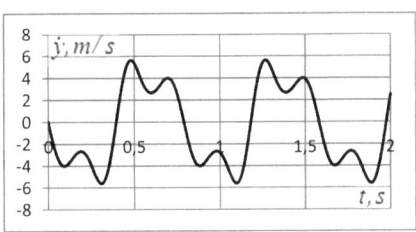

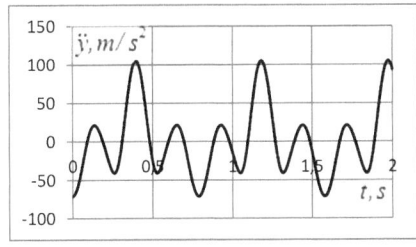

*Fig 2/10 Kinematical diagrams of oscillating mass point in case of big displacement ($0\,s \leq t \leq 2\,s$)*

Some remarks to Figs 2/8-11:

➢ In case of small displacements the oscillation is quasi harmonic. The characteristic of vertical component of (linear) spring force is quasi linear due to short trajectory. In initial position the spring is unloaded and after that it becomes compressed.

➢ In case of big displacements the motion of mass point differs strongly from harmonic oscillation. Illustrated component of spring force is absolutely nonlinear. In starting position of mass point ($y_o = 0{,}6\,m$) the spring is tensioned.

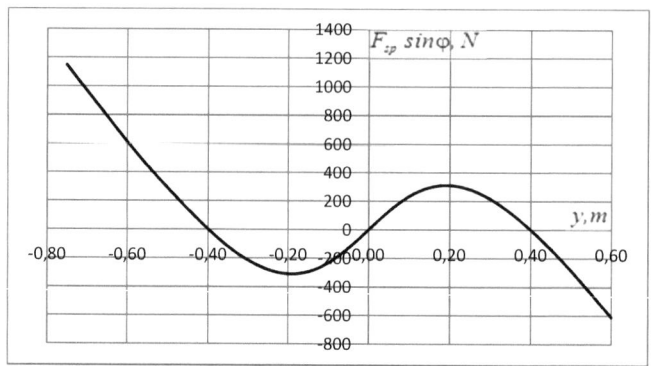

*Fig 2/11 Vertical component of spring force in function of position of mass point (big displacements)*

## 2.2. Natural frequencies of torsional vibration

### 2.2.1. Single degree-of-freedom oscillation

**Example 2/5** – In Fig 2/12 a propped beam can be seen. Its cross-section is round. There is a disc fastened to free end of the beam. Starting from rest position after rotation of $\varphi_{max}$ angular displacement the beam became prestressed. Released the disc the system oscillates torsional.

Calculate the natural frequency of torsional oscillating system for small angular displacements! Data: *l=400 mm, d=10 mm, D=200 mm, m=6 kg, G=80 GPa*.

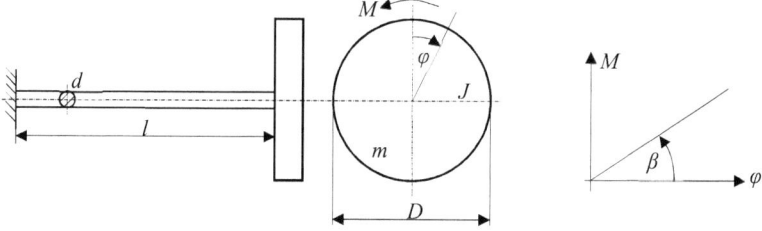

*Fig 2/12 Sketch of torsional oscillating system*

*Fig 2/13 Linear characteristic of torsional spring*

On the basis of knowledge of strength analysis when the length *l* and diameter *d* of the beam furthermore the torque *M* acting on it the angular displacement between its ends is

$$\varphi = \frac{l}{I_p G} M = \frac{32l}{d^4 \pi G} M \,, \qquad (2/11)$$

from which the spring stiffness (Fig 2/13)

$$s = tg\beta = \frac{M}{\varphi} = \frac{d^4 \pi G}{32l} \,, \qquad (2/12)$$

where $G$ is the modulus of elasticity in shear of the material of the beam.

Torque acting on from rest position deflected disc and the motion equation is

$$M = -s\varphi = J\ddot{\varphi}. \qquad (2/13)$$

Direction of the torque and the angular displacement are opposite. Differential-equation of oscillating system is

$$\ddot{\varphi} + \frac{s}{J}\varphi = 0 \,, \qquad (2/14)$$

where $J$ is the inertial moment of disc calculated for axis of rotation.

$$J = \frac{1}{2}m(\frac{D}{2})^2 = \frac{1}{2} \cdot 6kg \cdot (\frac{0.2m}{2})^2 = 0.03 kgm^2 \,. \qquad (2/15)$$

The structure of this differential-equation and differential-equation of harmonic oscillation are the same for this reason the natural frequency of system is

$$f = \frac{\omega}{2\pi} = \frac{1}{2\pi}\sqrt{\frac{s}{J}} = \frac{1}{2\pi}\sqrt{\frac{d^4 \pi G}{32lJ}} = \frac{1}{2\pi}\sqrt{\frac{(0.01m)^4 \pi 80 \cdot 10^9 \frac{N}{m^2} \cdot}{32 \cdot 0.4m \cdot 0.03 kgm^2}} = 12.88 s^{-1}. \qquad (2/16)$$

## 2.2.2. Multi degree-of-freedom torsional oscillation

Discs (gears, pulley disc, and so on) secured on shaft can oscillate rotating around the axis of shaft because of its elasticity. This is rotational or torsional oscillation. Two discs are fastened on ends of horizontal shaft as it can be seen in Fig 2/14.

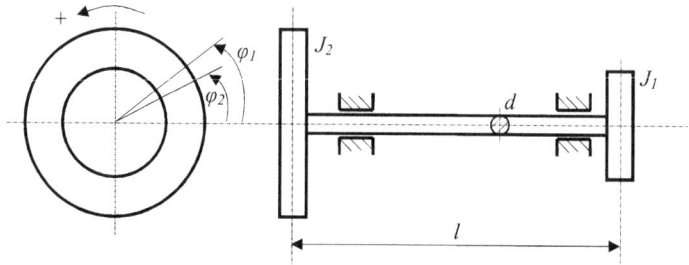

*Fig 2/14 Torsional oscillation of discs secured on elastic shaft*

From point of view of vibration theory this oscillating system and translational oscillating system consisting of two mass and one spring in between are the same. Orientations of discs are given by angular displacements $\varphi_1$ and $\varphi_2$. Torsional spring stiffness of the shaft is

$$s = I_p \frac{G}{l} = \frac{d^4 \pi}{32} \frac{G}{l}. \qquad (2/17)$$

Motion equations of two degree-of-freedom torsional oscillating system are

$$J_1 \ddot{\varphi}_1 = -s(\varphi_1 - \varphi_2), \quad J_2 \ddot{\varphi}_2 = s(\varphi_1 - \varphi_2), \qquad (2/18)$$

from which after rearrangement the

$$J_1 \ddot{\varphi}_1 + s\varphi_1 - s\varphi_2 = 0, \quad J_2 \ddot{\varphi}_2 - s\varphi_1 + s\varphi_2 = 0, \qquad (2/19)$$

homogenous differential equation-system can be obtained. Above equation-system can be written down in matrix form, i.e.

$$\mathbf{M\ddot{q}} + \mathbf{Sq} = 0, \qquad (2/20)$$

where $\mathbf{M}$ mass matrix and $\mathbf{S}$ spring stiffness matrix. In details,

$$\begin{bmatrix} J_1 & 0 \\ 0 & J_2 \end{bmatrix} \begin{bmatrix} \ddot{\varphi}_1 \\ \ddot{\varphi}_2 \end{bmatrix} + \begin{bmatrix} s & -s \\ -s & s \end{bmatrix} \begin{bmatrix} \varphi_1 \\ \varphi_2 \end{bmatrix} = \begin{bmatrix} 0 \\ 0 \end{bmatrix}. \tag{2/21}$$

In order to determine the natural frequencies of oscillating system the homogenous matrix differential equation has to be solved. Looking for the solution in form of

$$\varphi_1 = \varphi_{1o} \sin(\omega t + \alpha_1), \qquad \varphi_2 = \varphi_{2o} \sin(\omega t + \alpha_2), \tag{2/22}$$

from which

$$\dot{\varphi}_1 = \omega\varphi_{1o} \cos(\omega t + \alpha_1), \qquad \dot{\varphi}_2 = \omega\varphi_{2o} \cos(\omega t + \alpha_2),$$
$$\ddot{\varphi}_1 = -\omega^2\varphi_{1o} \sin(\omega t + \alpha_1), \qquad \ddot{\varphi}_2 = -\omega^2\varphi_{2o} \sin(\omega t + \alpha_2). \tag{2/23}$$

Substituting above kinematical equations into the homogenous matrix differential equation, after rearrangement the following equation-system can be obtained

$$\begin{bmatrix} -J_1\omega^2 + s & -s \\ -s & -J_2\omega^2 + s \end{bmatrix} \begin{bmatrix} \varphi_{1o} \\ \varphi_{2o} \end{bmatrix} = \begin{bmatrix} 0 \\ 0 \end{bmatrix}. \tag{2/24}$$

Trivial solution is $\varphi_{1o} = \varphi_{2o} = 0$ which means state of rest. In case of real solution the determinant of matrix equals to zero, i.e.

$$\begin{vmatrix} -J_1\omega^2 + s & -s \\ -s & -J_2\omega^2 + s \end{vmatrix} = 0. \tag{2/25}$$

After development of determinant and rearrangement,

$$J_1 J_2 \omega^4 - (J_1 + J_2)\omega^2 s = 0, \tag{2/26}$$

from where $\omega_{1,2} = 0$, respectively $\omega_{3,4} = \pm\sqrt{\dfrac{(J_1 + J_2)s}{J_1 J_2}}$.

**Example 2/6** – Calculate the natural angle frequencies of torsional oscillating system (Fig 2/14) and plot its kinematical functions! Data (see Fig 2/14): $J_1 = 2.5\ kgm^2$, $J_2 = 4\ kgm^2$, $d = 10\ mm$, $l = 1\ m$, $G = 80\ GPa$.

Torsional spring stiffness of the shaft is

$$s = I_p \frac{G}{l} = \frac{d^4\pi}{32} \frac{G}{l} = \frac{(0.01m)^4\pi}{32} \frac{80 \cdot 10^9 \frac{N}{m^2}}{1m} = 78.54 \frac{Nm}{rad}. \quad (2/27)$$

The meaning of root $\omega_{12} = 0$ is merely the rotation or the possibility of rotation of discs at a constant angular velocity. From physical point of view the important root is the positive natural angle frequency, i.e.

$$\omega_{34} = \sqrt{\frac{(J_1 + J_2)s}{J_1 J_2}} = \sqrt{\frac{(2.5 + 4)kgm^2 \cdot 78.54 \frac{Nm}{rad}}{2.5kgm^2 \cdot 4kgm^2}} = 7.145s^{-1}. \quad (2/28)$$

Differential-equation system (2/18) can be solved applying algorithms in MS Excel can be seen in table on next page.

Kinematical functions of the oscillation using above given data in Fig 2/16 can be shown. Initial conditions: $\varphi_{1o} = 0\ rad$, $\dot{\varphi}_{1o} = 0\ rads^{-1}$, $\varphi_{2o} = 0\ rad$, $\dot{\varphi}_{2o} = 2\ rads^{-1}$. It can be noticed discs oscillate at calculated natural frequency in opposite phase.

When among initial conditions the initial angular velocity of disc denoted by $1$ is different from zero, discs rotate at uniform angular velocity and oscillate simultaneously (Fig 2/16).

The phenomena of resonance can be studied in case of torsional oscillation as well (Fig 2/15). Disc denoted by $1$ is excited by sinusoidal torsional moment according to function of $M = M_{1o} \cos \omega_g t$, where $M_{1o}$ is the amplitude of torsional moment and $\omega_g$ is the above calculated natural angle frequency.

Without damping the angular displacements of discs are getting bigger and bigger. (Used data are the same.)

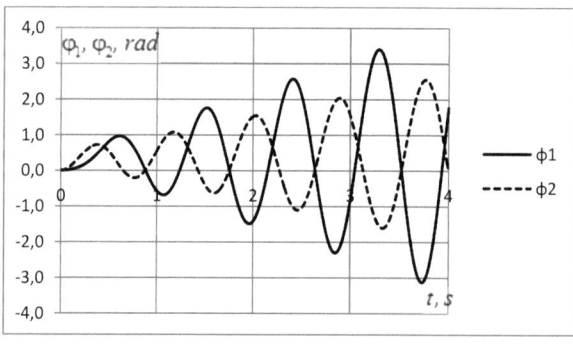

*Fig 2/15 Phenomena of resonance in case of torsional oscillation ($M_{1o}=10\ Nm$, $0\ s \leq t \leq 4\ s$)*

Applied algorithms for solving differential-equation system (Example 2/6)

| $t$ | $\ddot{\varphi}_1(\varphi_1,\varphi_2)$ | $\dot{\varphi}_1$ | $\varphi_1$ | $\ddot{\varphi}_2(\varphi_1,\varphi_2)$ | $\dot{\varphi}_2$ | $\varphi_2$ |
|---|---|---|---|---|---|---|
| $t_o$ | $\ddot{\varphi}_{1o}(\varphi_{1o},\varphi_{2o})$ | $\dot{\varphi}_{1o}$ | $\varphi_{1o}$ | $\ddot{\varphi}_{2o}(\varphi_{1o},\varphi_{2o})$ | $\dot{\varphi}_{2o}$ | $\varphi_{2o}$ |
| $t_1$ | $\ddot{\varphi}_{11}(\varphi_{11},\varphi_{21})$ | $\dot{\varphi}_{11}=\dot{\varphi}_{1o}+\ddot{\varphi}_{11}(t_1-t_o)$ | $\varphi_{11}=\varphi_{1o}+\dot{\varphi}_{11}(t_1-t_o)$ | $\ddot{\varphi}_{21}(\varphi_{11},\varphi_{21})$ | $\dot{\varphi}_{21}=\dot{\varphi}_{2o}+\ddot{\varphi}_{21}(t_1-t_o)$ | $\varphi_{21}=\varphi_{2o}+\dot{\varphi}_{21}(t_1-t_o)$ |
| $t_2$ | $\ddot{\varphi}_{12}(\varphi_{12},\varphi_{22})$ | $\dot{\varphi}_{12}=\dot{\varphi}_{11}+\ddot{\varphi}_{12}(t_2-t_1)$ | $\varphi_{12}=\varphi_{11}+\dot{\varphi}_{12}(t_2-t_1)$ | $\ddot{\varphi}_{22}(\varphi_{12},\varphi_{22})$ | $\dot{\varphi}_{22}=\dot{\varphi}_{21}+\ddot{\varphi}_{22}(t_2-t_1)$ | $\varphi_{22}=\varphi_{21}+\dot{\varphi}_{22}(t_2-t_1)$ |
| $t_3$ | .... | .... | .... | .... | .... | .... |

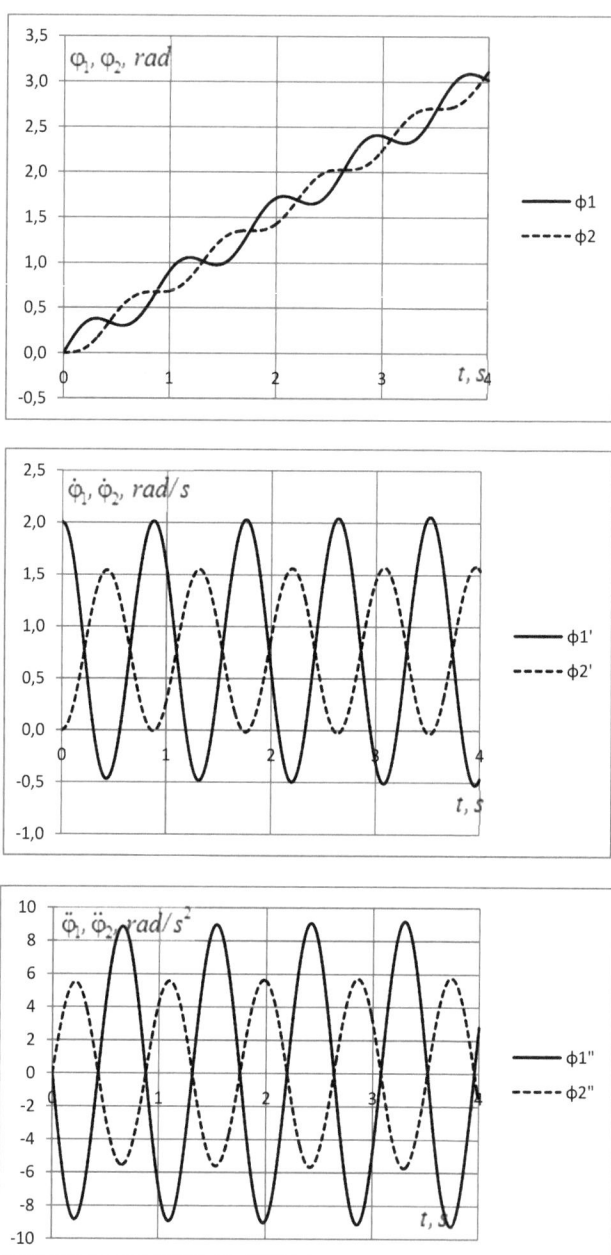

*Fig 2/16 Kinematical functions of torsional oscillation of two discs secured on ends of elastic shaft (0 s ≤ t ≤ 4 s)*

**Example 2/6** – There are three secured discs on horizontal shaft (Fig 2/17). Orientation of them are described by angular displacements, $\varphi_1$, $\varphi_2$ and $\varphi_3$.

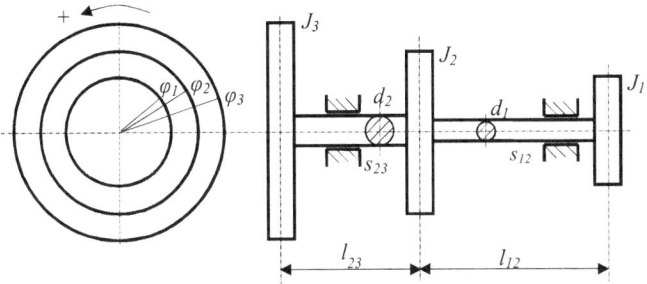

*Fig 2/17 Torsional oscillation of discs secured on elastic shaft*

It is about undamped free torsional oscillation. Torsional spring stiffness of each part of the shaft is

$$s_{12} = I_{p1} \frac{G}{l_{12}} = \frac{d_1^4 \pi}{32} \frac{G}{l_{12}}, \qquad s_{23} = I_{p2} \frac{G}{l_{23}} = \frac{d_2^4 \pi}{32} \frac{G}{l_{23}}. \qquad (2/29)$$

Motion equations of three degree-of-freedom torsional oscillating system are

$$J_1 \ddot{\varphi}_1 = -s_{12}(\varphi_1 - \varphi_2),$$

$$J_2 \ddot{\varphi}_2 = s_{12}(\varphi_1 - \varphi_2) - s_{23}(\varphi_2 - \varphi_3), \qquad (2/30)$$

$$J_3 \ddot{\varphi}_3 = s_{23}(\varphi_2 - \varphi_3),$$

from which after rearrangement the following homogenous matrix differential-equation can be obtained, i.e.

$$\begin{bmatrix} J_1 & 0 & 0 \\ 0 & J_2 & 0 \\ 0 & 0 & J_3 \end{bmatrix} \begin{bmatrix} \ddot{\varphi}_1 \\ \ddot{\varphi}_2 \\ \ddot{\varphi}_3 \end{bmatrix} + \begin{bmatrix} s_{12} & -s_{12} & 0 \\ -s_{12} & s_{12}+s_{23} & -s_{23} \\ 0 & -s_{23} & s_{23} \end{bmatrix} \begin{bmatrix} \varphi_1 \\ \varphi_2 \\ \varphi_3 \end{bmatrix} = \begin{bmatrix} 0 \\ 0 \\ 0 \end{bmatrix}. \qquad (2/31)$$

To determine natural frequencies looking for the solution again in form of

$$\varphi_1 = \varphi_{1o} \sin(\omega t + \alpha_1), \quad \varphi_2 = \varphi_{2o} \sin(\omega t + \alpha_2),$$

$$\varphi_3 = \varphi_{3o} \sin(\omega t + \alpha_3). \qquad (2/32)$$

47

Substituting above kinematical functions into the homogenous matrix differential-equation, after rearrangement the

$$
\begin{bmatrix}
-J_1\omega^2 + s_{12} & -s_{12} & 0 \\
-s_{12} & -J_2\omega^2 + s_{12} + s_{23} & -s_{23} \\
0 & -s_{23} & -J_3\omega^2 + s_{23}
\end{bmatrix}
\begin{bmatrix}
\varphi_{1o} \\
\varphi_{2o} \\
\varphi_{3o}
\end{bmatrix}
=
\begin{bmatrix}
0 \\
0 \\
0
\end{bmatrix}
\tag{2/33}
$$

equation-system can be obtained. Solution different from trivial one can be reached when the determinant of matrix equals to zero, i.e.

$$
\begin{vmatrix}
-J_1\omega^2 + s_{12} & -s_{12} & 0 \\
-s_{12} & -J_2\omega^2 + s_{12} + s_{23} & -s_{23} \\
0 & -s_{23} & -J_3\omega^2 + s_{23}
\end{vmatrix}
= 0 .
\tag{2/34}
$$

After development of determinant and rearrangement the characteristic equation of oscillating system is

$$
\omega^2\left[-J_1J_2J_3\omega^4 + (J_1J_2s_{23} + J_1J_3(s_{12}+s_{23}) + J_2J_3s_{12})\omega^2 - s_{12}s_{23}(J_1+J_2+J_3)\right] = 0,
\tag{2/35}
$$

from where $\omega_{1,2} = 0$, respectively

$$
\omega_{3,4}^2 = \frac{-[J_1J_2s_{23} + J_1J_3(s_{12}+s_{23}) + J_2J_3s_{12}]}{2J_1J_2} \pm
$$
$$
\pm \frac{\sqrt{[J_1J_2s_{23} + J_1J_3(s_{12}+s_{23}) + J_2J_3s_{12}]^2 - 4J_1J_2J_3s_{12}s_{23}(J_1+J_2+J_3)}}{2J_1J_2} .
\tag{2/36}
$$

When $J_1=2\ kgm^2$, $J_2=3\ kgm^2$, $J_3=4\ kgm^2$, $d_1=10\ mm$, $d_2=12\ mm$, $l_{12}=l_{23}=0.8\ m$, $G=80\ GPa$, from physical point of view the important natural angle frequencies are $\omega_1=0\ s^{-1}$, $\omega_2=7.05\ s^{-1}$, $\omega_3=12.28\ s^{-1}$.

In Fig 2/18 the superposition of uniform rotation and torsional oscillation can be seen. Initial conditions are $\varphi_{1o} = 0rad$, $\dot{\varphi}_{1o} = 2rads^{-1}$, $\varphi_{2o} = 0rad$, $\dot{\varphi}_{2o} = 0rads^{-1}$, $\varphi_{3o} = 0rad$, $\dot{\varphi}_{3o} = 0rads^{-1}$.

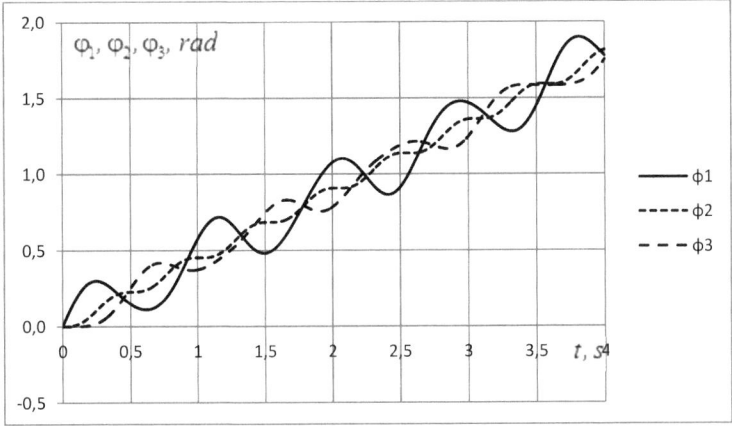

*Fig 2/18 Angular displacement-time functions of torsional rotation of three discs fastened on elastic shaft (0 s ≤ t ≤ 4 s)*

## 2.3. Critical shaft speed

Perfect balance of discs secured to elastic shafts can be never accomplished for this reason the rotating shaft becomes bent due to inertial forces. At critical shaft speed (or speeds) the unbalance and the deformation of the shaft become infinite.

It is evident for the constructor to create the mechanical system consisting of elastic shaft and disc(s) that the critical shaft speed and working speed should be different at needed degree of safety.

Simplest mechanical model of disc secured on elastic shaft can be seen in Fig 2/19. Length of horizontal shaft supported in bearings denoted by $l$.

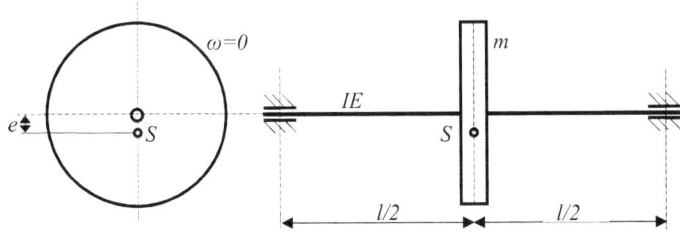

*Fig 2/19 Model of disc secured on elastic shaft*

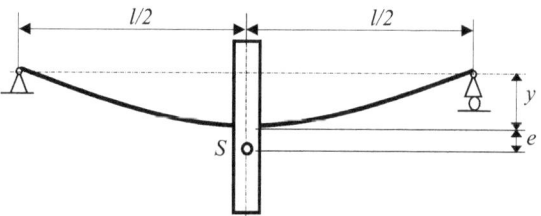

*Fig 2/20 Rotating deformed shaft ($\omega \leq \sqrt{2}\omega_k$)*

Disc is located at the center of the shaft. The disc is static unbalanced, distance between center of mass and axis of rotation is denoted by $e$. When shaft speed is less than critical one the shaft rounds away for this reason the center of mass goes along a circle. Its radius equals to $r = y + e$ (Fig 2/20). While it is rotating the inertial force equals to spring force of bent shaft, i.e.

$$m(y+e)\omega^2 = sy , \qquad (2/37)$$

where spring stiffness of the shaft from strength of materials is

$$s = \frac{48IE}{l^3} . \qquad (2/38)$$

After rearrangement

$$y = \frac{e\omega^2}{\dfrac{s}{m} - \omega^2} . \qquad (2/39)$$

When denominator tends to zero the amplitude becomes infinite for this reason the angular velocity is

$$\omega = \sqrt{\frac{s}{m}} = \omega_k , \qquad (2/40)$$

and to its adequate shaft speed is called critical shaft speed.

This result is formally similar to solution obtained at bending vibration where the load of the beam is alternating bending. If gravitational force of disc can be neglected, the load of the shaft is static bending near critical speed in spite of the fact that it is a vibration generator across built-in bearings.

On the basis of obtained solution it can be established the following: if the working speed is less than critical shaft speed the amplitude can be kept in a safe interval. Rewrite the obtained result,

$$r = y + e = \frac{e\omega^2}{\dfrac{s}{m} - \omega^2} + e = \frac{e}{1 - (\dfrac{\omega}{\omega_k})^2},$$  (2/41)

respectively

$$\frac{r}{e} = \frac{1}{1 - (\dfrac{\omega}{\omega_k})^2}.$$  (2/42)

The equation is plotted in Fig 2/21.

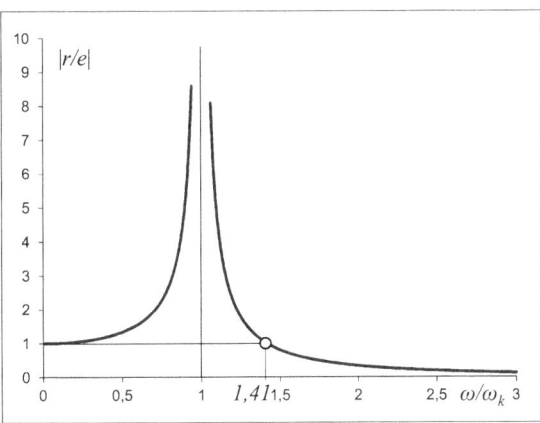

Fig 2/21 The amplitude of disc in function of working speed of shaft

After starting the amplitude of disc is getting higher and higher, later after the critical interval tends to zero.

When $\frac{\omega}{\omega_k}\rangle\sqrt{2}$, $\left|\frac{r}{e}\right|\langle 1$, i.e. the center of mass of disc moving along a circle gets closer and closer to axis of rotation than the initial eccentricity. It is called self-balancing. At critical speed shaft the center of mass gets to opposite side of axis of rotation.

In order to determine the spring stiffness of the shaft and finally the critical shaft speed in other cases it has to be known deformations caused by forces acting on the shaft. Methods to do them can be found in strength of materials for example Betti theorem. Slope-deflection equations of elastic prismatic beams can be seen in Fig 2/22-23.

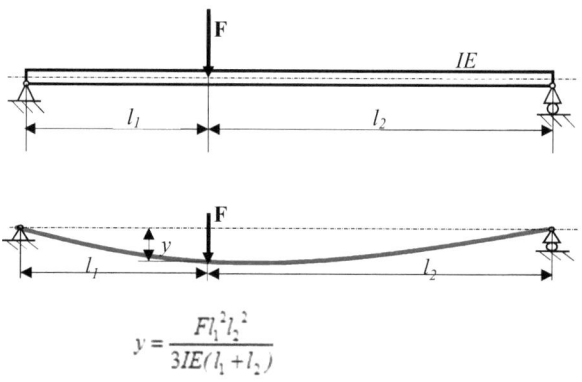

$$y = \frac{Fl_1^2 l_2^2}{3IE(l_1+l_2)}$$

*Fig 2/22 Elastic deflection of point of application of force denoted by* **F**

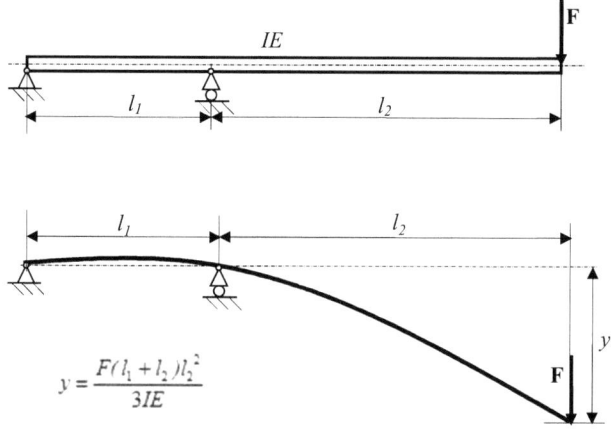

$$y = \frac{F(l_1+l_2)l_2^2}{3IE}$$

*Fig 2/23 Elastic deflection of point of application of force denoted by* **F**

52

**Example 2/7** – There is a disc secured on elastic shaft as it can be seen in Fig 2/24. Calculate the critical speed shaft!

Data: length of the shaft $l=2$ $m$, circular cross-section, diameter $d=15$ $mm$, Young's modulus of its material $E=210$ $GPa$, diameter of the disc $D=400$ $mm$, thickness $v=40$ $mm$, density of its material $\rho=7850$ $kg/m^3$, $l_1=0.5$ $m$, $l_2=1.5$ $m$.

Using the adequate slope-deflection equation from Fig 2/22 the spring stiffness of the shaft is

$$s = \frac{F}{y} = \frac{3IE(l_1+l_2)}{l_1^2 l_2^2} = \frac{3d^4\pi \cdot E(l_1+l_2)}{64 \cdot l_1^2 l_2^2} =$$
$$= \frac{3 \cdot (0.015m)^4 \pi \cdot 210 \cdot 10^9 \, N/m^2 \cdot (0.5m+1.5m)}{64 \cdot (0.5m)^2 (1.5m)^2} = 5566.52 \frac{N}{m}. \qquad (2/43)$$

The mass of the disc is

$$m = \rho \frac{D^2 \pi}{4} v = 7850 \frac{kg}{m^3} \frac{(0.4m)^2 \pi}{4} 0.04m = 39.46 kg, \qquad (2/44)$$

and finally the critical speed shaft is

$$n_k = \frac{1}{2\pi}\sqrt{\frac{s}{m}} = \frac{1}{2\pi}\sqrt{\frac{5566.52 \dfrac{N}{m}}{39.46 kg}} = 1.89 s^{-1} = 113.42 \, min^{-1}. \qquad (2/45)$$

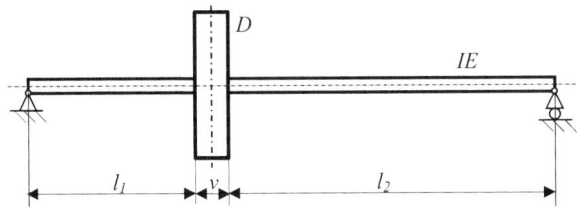

*Fig 2/24 Disc secured on elastic shaft*

**Example 2/8** – There is a disc secured on vertical shaft of mixing equipment (Fig 2/25). Calculate the critical speed shaft!

Data: circular cross-section of the shaft, diameter $d=20$ $mm$, Young's modulus of its material $E=210$ $GPa$, diameter of the disc $D=400$ $mm$, thickness $v=40$ $mm$, density of its material $\rho=7850\ kg/m^3$, distance between supports and the length of the bracket $l_1=0.5\ m$, $l_2=1.0\ m$.

Mass of the disc from previous example $m=39.46\ kg$. The slope-deflection equation can be found in Fig 2/23 and applying it the spring stiffness is

$$
\begin{aligned}
s = \frac{F}{y} &= \frac{3IE}{(l_1+l_2)l_2^{\ 2}} = \frac{3d^4\pi\cdot E}{64\cdot(l_1+l_2)l_2^{\ 2}} = \\
&= \frac{3\cdot(0.015m)^4\pi\cdot210\cdot10^9\ N/m^2}{64\cdot(0.5m+1m)\cdot(1m)^2} = 3298.68\frac{N}{m},
\end{aligned}
\tag{2/46}
$$

and finally the critical speed shaft,

$$
n_k = \frac{1}{2\pi}\sqrt{\frac{s}{m}} = \frac{1}{2\pi}\sqrt{\frac{3298.68\dfrac{N}{m}}{39.46kg}} = 1.46s^{-1} = 87.31min^{-1}.
\tag{2/47}
$$

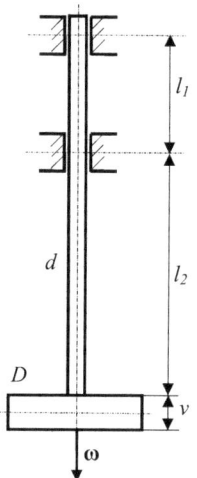

*Fig 2/25 Sketch of vertical shaft of mixing equipment*

Gyroscopic effect of discs could be taken into consideration but it is observable that critical speed shaft is not influenced by gyroscopic effect significantly.

# References

JALON, J. G., BAYO, E.: Kinematic and dynamic simulation of multibody systems –
The real-Time Challenge. 1994. Springer-Verlag, New York. ISBN 0-387-94096-0

SHIGLEY, J. E., UICKER, J. J.: Theory of machines and mechanisms. McGraw-Hill
Series in Mechanical Engineering, 1981, ISBN 0-07-056884-7

DUKKIPATI, R. V.: Mechanism and Machine Theory. New Age International Ltd.,
1989, ISBN 81-224-0426-X

BIRO, I.: Mechanical vibrations. University of Szeged, Faculty of Engineering, Szeged, Hungary, 2014, ISBN 978-963-306-288-3

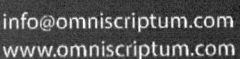